Real Science-4-Kids

SCIENCE

Book 1

Rebecca Woodbury, Ph.D., M.Ed. **RS4K**

RS4K Real Science-4-Kids

Illustrations: Janet Moneymaker

Science Book 1 Student Textbook (hardcover)
ISBN 978-1-941181-76-8

Published by Gravitas Publications Inc.
Real Science-4-Kids®
www.gravitaspublications.com

G GRAVITAS PUBLICATIONS INC.

► Table of Contents

Physics

Geology

Conclusion

What Is SCIENCE?

1.1 Introduction

Have you noticed that ice always floats in a cup of cold water? Do you think about how a flower blooms or why leaves are green? Do you wonder what happens to the Sun at night? Have you thought about how wings help a bird fly?

White Southern Magnolia Flower

All of these questions are about how things work. **Science** provides us with a method for studying how things work.

Science has several different parts. The first part of science is making **observations** and asking questions.

The second part of science is thinking of ways to answer these questions. To answer questions, **scientists** do **experiments** to test ideas and collect information. Testing ideas makes science different from all other types of study.

The last part of science is comparing ideas with other scientists and sometimes even arguing about the answers to the questions!

In summary, scientists observe and ask questions, do experiments, and compare their experiments to those of other scientists. This is what science is all about!

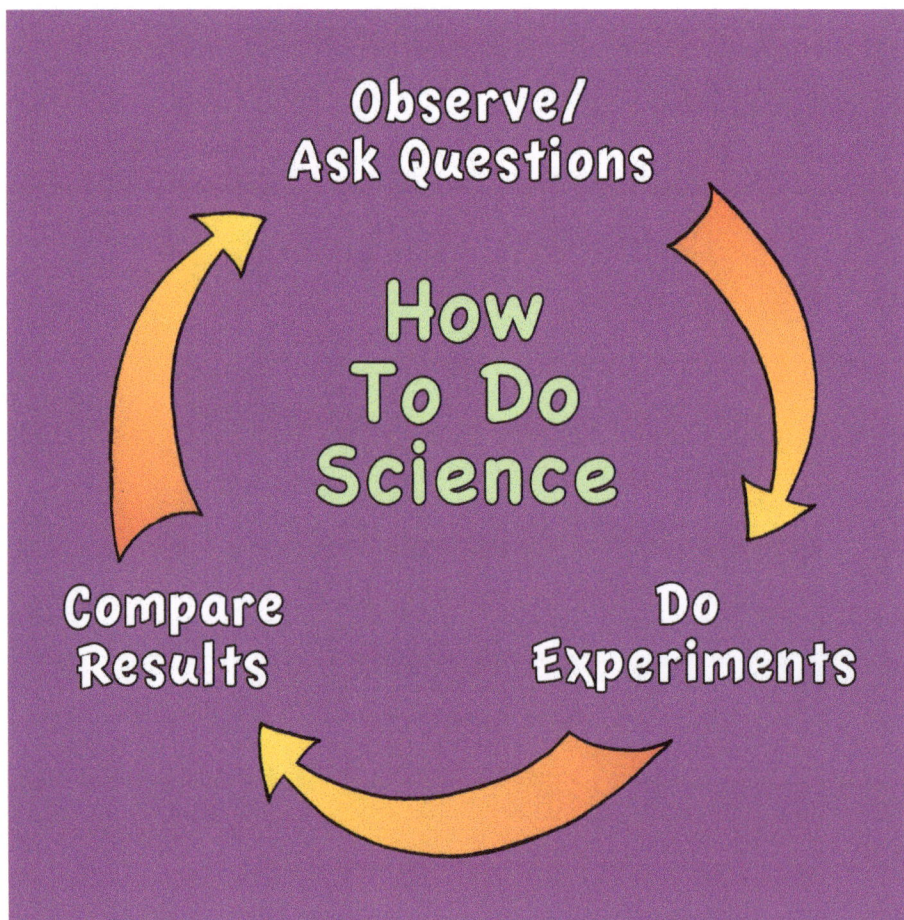

Observe/
Ask Questions

How
To Do
Science

Compare
Results

Do
Experiments

1.2 History of Science

A very long time ago students did not study "science." They might have studied Latin, music, or poetry, but there were no books on science and there were no scientists. So where did science come from?

The history of science is long. It's not easy to say exactly where and when science started.

We know that ancient people learned how to grow food, studied the stars, knew about planets, and even understood enough chemistry to refine gold and silver so they could make coins and jewelry. Even though ancient people didn't formally study science, they contributed ideas and knowledge that helped lead to what we now call science.

WHERE'S THE SCIENCE?

The science that we study now is like a story that has been written and rewritten many times. Many people over many thousands of years have contributed to the story of science.

Today, the way we explain how things work is very different from the way ancient people explained how things work. Over time scientists have changed the story. For example, people used to believe that the Sun moved around Earth. But today we know that Earth moves around the Sun. By helping us understand gravity, motion, and how planets move, scientific discoveries have changed the story of how Earth moves.

El Caracol, The Observatory at Chichen Itza used by the ancient Mayan people to study objects in the sky

It is important to keep in mind that stories for how things work continue to change. Even today we don't know everything about the world around us or everything about how things work.

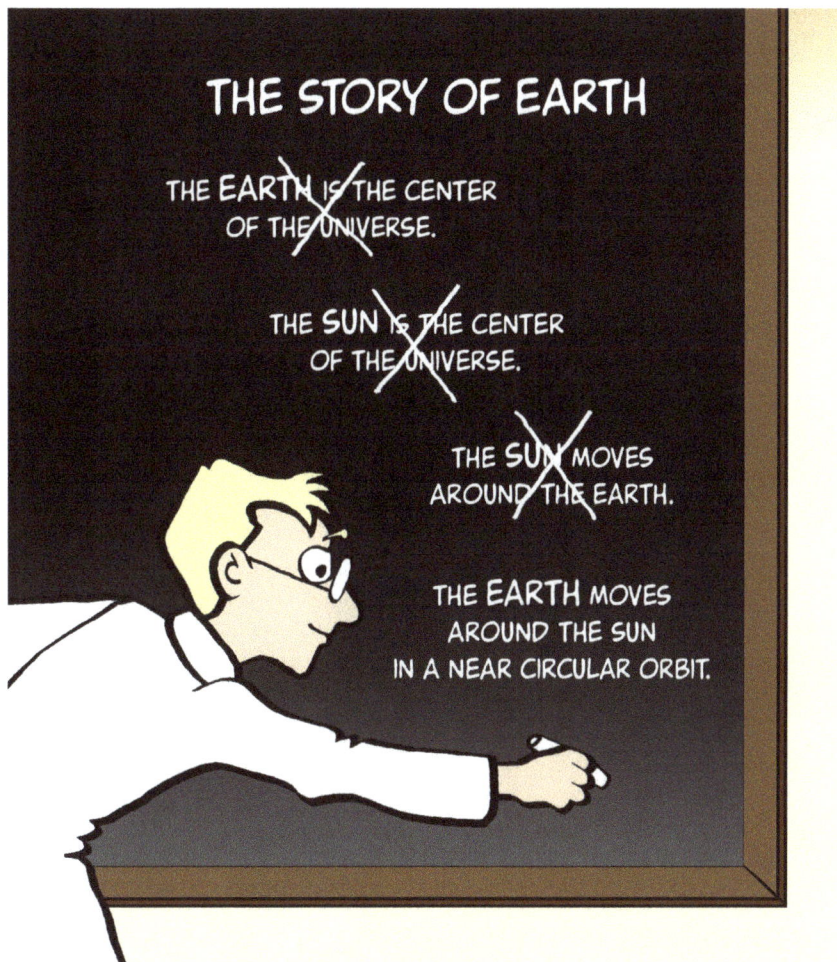

As scientists learn more and more, they provide better and better explanations of how things work. These explanations may either change or confirm ideas that we believe are true. Or they may lead to the development of surprising new ideas. This is how new scientific discoveries are made and how science helps us understand the world around us.

1.3 Building Blocks of Science

Modern science is a combination of different scientific subjects. A **scientific subject** is a specific area of science that someone studies, explores and researches.

Although there are many different scientific subjects, five subjects make up the foundation of all science: **chemistry, biology, physics, geology, and astronomy.**

It is easy to think about studying science if we think about the different scientific subjects as building blocks that fit together and rely on each other.

To understand science a student needs to study at least some of the important topics in each of these five scientific subjects. For example, atoms are studied in the chemistry building block, force in the physics building block, and plants in the biology building block. These are three different essential topics in three different building block subjects.

Often, a scientific topic in a particular scientific building block cannot be understood without knowing about topics in other building blocks. For example, in astronomy to understand how planets move around the Sun, a student must learn about gravity. Gravity is a topic typically covered in physics. So physics is a building block necessary for understanding astronomy.

Another example is understanding how plants make food. Both chemistry and physics are required for an understanding of the biology of how plants make food.

Chemistry and physics are both building blocks that are essential for understanding the building block of biology.

In this book you will be introduced to many different science topics. You will learn about atoms in chemistry, cells in biology, energy in physics, the Earth in geology, and Earth's orbit in astronomy. You will see how the five scientific building block subjects fit together to provide an understanding of different science topics.

When the five essential building blocks of science are put together, a solid foundation for understanding science is formed.

1.4 The Scientific Method

To do science, a scientist loosely follows a certain set of steps. These steps are called the **scientific method.**

▶ The first and most important step is to make good **observations** about everything you see, hear, and touch.

A scientist doesn't just think about objects but looks at objects in detail. How heavy is it? How hot or cold is it? What color is it? By asking questions, the scientist is making observations.

Making good observations is the first and most essential step in the scientific method. A scientist can't study something without having carefully observed it first.

▶ The second step in the scientific method is asking a question about what has been observed and then turning that question into a **hypothesis**.

A hypothesis is a statement about something. For example, a question might be, "How heavy is an elephant?" Turning that question into a hypothesis, it becomes, "An elephant is heavier than a kangaroo."

▶ The third step in the scientific method is creating an **experiment** to prove or disprove the hypothesis.

Is an elephant heavier than a kangaroo? The scientist doesn't make any assumptions about the answer but performs an experiment to find out. A big scale might be used to measure how much each animal weighs. The scientist might weigh the kangaroo first or the elephant first and then the kangaroo. The scientist designs the experiment to prove or disprove the hypothesis.

By measuring how much an elephant weighs and how much a kangaroo weighs, the scientist can then move to the next step of the scientific method.

▶ The fourth step of the scientific method is **collecting results** by writing or recording data.

What was the outcome of the experiment? What were the actual numbers observed on the scale?

The results of the experiment that was performed are written down. Writing down the numbers means that the scientist doesn't have to guess about the results of the experiment. It is important that the information is recorded exactly as it is observed and that no guesses are made about what the answer might be. By not guessing what the answer might be, a scientist remains open to new discoveries.

Finally, the scientist is ready for the last step of the scientific method.

▶ The last step of the scientific method is to **draw a conclusion**.

What did the information show? Was the elephant heavier than the kangaroo or was the kangaroo heavier than the elephant?

The conclusion comes from the results of the experiment. Based on the results, the scientist draws a conclusion and shows that the hypothesis has been either proven or not proven. Conclusions can change when new data are collected.

1.5 Summary

• **Science** is a way to study how things work.

• **Scientists** make observations, ask questions, do experiments, and compare their experiments with other scientists.

• There are five scientific subjects that make up the essentials of all science: **chemistry**, **biology**, **physics**, **geology**, and **astronomy**.

• To do science, a scientist follows a certain set of steps. These steps are called the **scientific method**.

I can't wait to learn science!

1.6 Vocabulary

Basic Words

science   (SIY-uhns)

method   (MEH-thuhd)

physics   (FIH-zihks)

Challenge Words

astronomy   (uh-STRAH-nuh-mee)

biology   (biy-AH-luh-jee)

chemistry   (KEH-muh-stree)

geology   (jee-AH-luh-jee)

observation   (ahb-suhr-VAY-shuhn)

What Is CHEMISTRY?

2.1 Introduction

In Chapter 1 you learned that science is a way to study how the world works. You also learned that science is made up of five different scientific subjects: **chemistry**, **biology**, **physics**, **astronomy**, and **geology**. These five subjects are the building blocks for all science.

In this chapter we will explore the building block called **chemistry**. Chemistry investigates what physical things are made of and the ways in which they change. Everything you can see, smell, touch, taste, or hear is made of "something." Scientists call this "something" **matter**. Chemistry is the study of matter.

2.2 History of Chemistry

A long time ago, students didn't study chemistry in school. Today, chemistry is an important part of all science, and students all over the globe study chemistry.

But where did we get chemistry?

Without knowing about chemistry, ancient people still did lots of chemistry. Ancient people used **chemical processes** every day.

▶ A chemical process is any activity that involves some kind of chemistry.

For example, when ancient people treated animal furs so that they could be used as clothing, they were doing chemistry.

I WONDER WHAT CHEMICAL REACTIONS THOSE ARE?

When ancient people created paints to draw pictures on walls, they were doing chemistry.

And when ancient people made coins or jewelry from metal, they were doing chemistry.

Two sides of an ancient Greek coin with images of Alexander the Great

Ancient people did a lot of **experimenting** to learn about the world around them. Experimenting just means testing ideas and observing what happens.

For example, by experimenting, ancient people learned whether or not something would burn or melt. They learned if something tasted sweet or bitter. They learned that some things change color when added to other things. They even learned that some plants could be used as medicine. All of this is chemistry.

However, people didn't start thinking about chemistry as a science until the late 16th century. Around this same time, other scientific subjects, such as physics and astronomy, were also being developed. Also at this time, there were lots of people who did not do any experimenting at all, but just thought about things.

Thinking about how things work and doing experiments are two different ways to learn about the world.

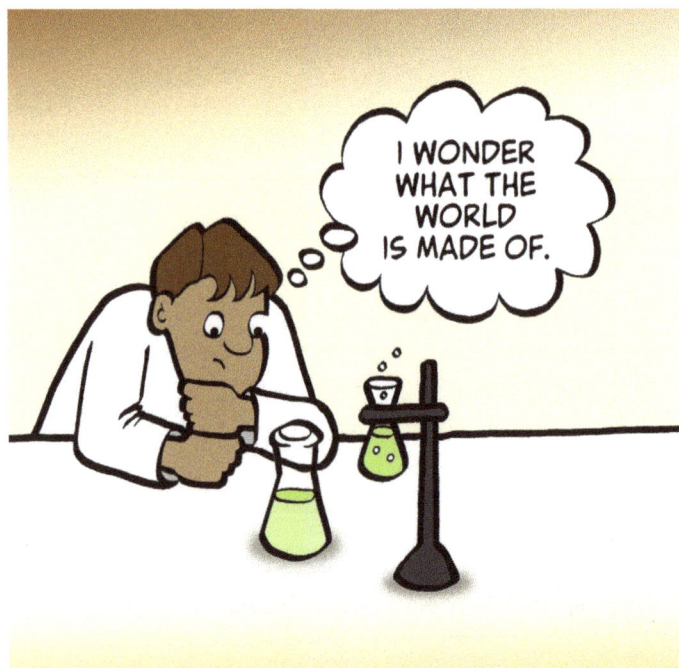

In the late 16th century, thinkers began thinking about experimenting. Eventually thinking and doing were blended together, and these two ways of studying the world turned into what we call chemistry.

2.3 Modern Chemistry

Thinking about how the world works and doing experiments to observe how the world works are both important to modern chemistry.

We will see in Chapter 3 that matter is made of **atoms**. But atoms are so small that they can't be seen with our eyes. If we can't see them with our eyes, how do we know they really exist? And if we can't see them with our eyes, how do we know what they look like?

Modern chemists use advanced technology and do experiments to show that atoms exist. From these experiments they can learn a great deal about atoms. For example, by doing an experiment a chemist might find out how large an atom is, how fast it can move, or how heavy it is.

Modern chemists also think about atoms and make models to explain how atoms behave. A **model** is a good guess about how something works. Models can be created as writing, drawings, or computer programs, or they can be built from different materials. A model may not show every detail and models don't show everything, but good models help chemists understand how things might work.

In Chapter 3 we use a model for atoms. Each atom is drawn with a face and arms. Atoms are also assigned a color. But real atoms don't have arms or faces and they don't have colors but that's OK. The models of atoms used in Chapter 3 help us understand how atoms work.

2.4 Everyday Chemistry

You use chemistry every day, but you probably don't know it.

▶ When you brush your teeth in the morning, chemicals in the toothpaste help clean your mouth. **This is chemistry**.

▶ If you have a tummy ache, you might take medicine to make you feel better. **This is chemistry.**

▶ If you want ice cream, your big sister fills the car with gas to drive you to the ice cream store. **This is chemistry**.

▶ If you want to paint a picture you might use watercolors. **This is chemistry.**

2.5 Summary

- **Chemistry** is the study of **matter**.

- **Chemistry** is thinking about how things work by making **models** and by doing **experiments**.

- **Chemistry** happens every day, and we use **chemistry** in many different ways.

2.6 Vocabulary

Basic Words

atom (AA-tuhm)

matter (MAA-tuhr)

Challenge Words

chemistry (KEH-muh-stree)

experiment (ihk-SPEHR-uh-muhnt)

I love to learn science words!

What Are ATOMS?

3.1 Atoms

Have you ever wondered if the Moon is made of green cheese?

Have you ever thought the clouds might be made of cotton candy?

Have you ever wanted to know what makes carrots orange...

or peas green?

Have you ever wondered why brussels sprouts couldn't taste more like sweet cherries, or asparagus taste more like candy canes?

Everything around us has a different shape or flavor or color because everything around us is put together with different combinations of atoms.

▶ An **atom** is the smallest unit that makes up matter. Atoms can combine with one another in different groupings to make different types of matter.

The Moon is not made of green cheese. It is made of the kinds of atoms that are found in rocks and dirt.

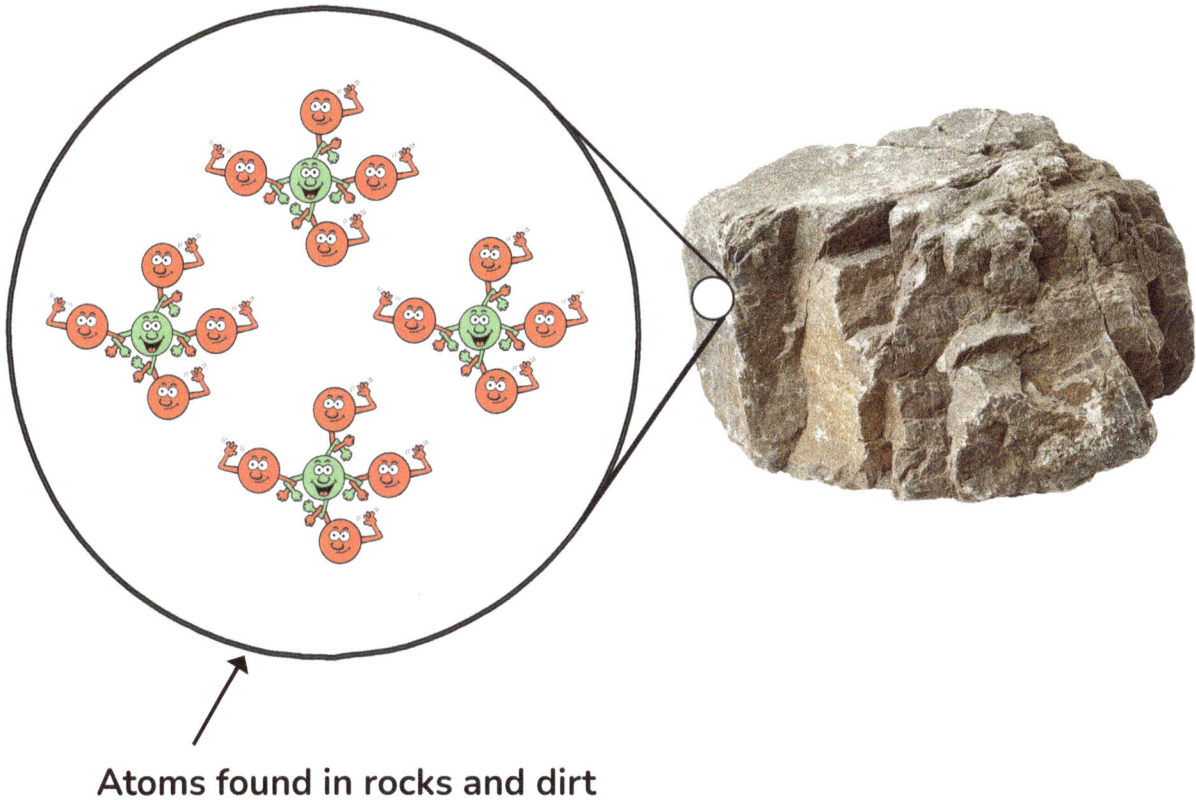

Atoms found in rocks and dirt

Clouds are not made of cotton candy, but of the kinds of atoms that are found in air and water.

Atoms inside clouds

Carrots are orange because their atoms are arranged in a way that makes them orange --but the atoms themselves are not orange.

Peas are green because their atoms are arranged in a way that makes them green--but the atoms themselves are not green.

Brussels sprouts don't taste sweet like cherries or candy canes because the atoms inside brussels sprouts are not arranged in a way that makes them sweet.

3.2 Types of Atoms

There are over 100 different atoms. **Carbon**, **oxygen**, and **nitrogen** are the names of a few of these atoms.

hydrogen

nitrogen

Different atoms

carbon

oxygen

chlorine

sodium

phosphorus

▶ Atoms are very tiny. They are so small that you can't see them with your eyes. Atoms don't get bigger when things get bigger. Atoms hook together to make big things (see Chapter 4).

Even though we can't see atoms with our eyes, we can make **models** of them. Remember from Chapter 2 that a model may not be exactly true but it can help chemists understand how things work. We can make models of atoms by drawing them as dots or little balls or by giving atoms colors and shapes.

In this book, atoms are modeled as balls with "arms" to help show how atoms hook to other atoms. Even though atoms don't really have arms, it's a great way to think about how atoms hook together. Also, atoms are given colors to help tell them apart, but in reality atoms don't have color.

A simple model of a silicon atom
shown with four "arms" and given the color blue

3.3 Atoms Are Similar

All atoms are made of **protons**, **neutrons**, and **electrons**.

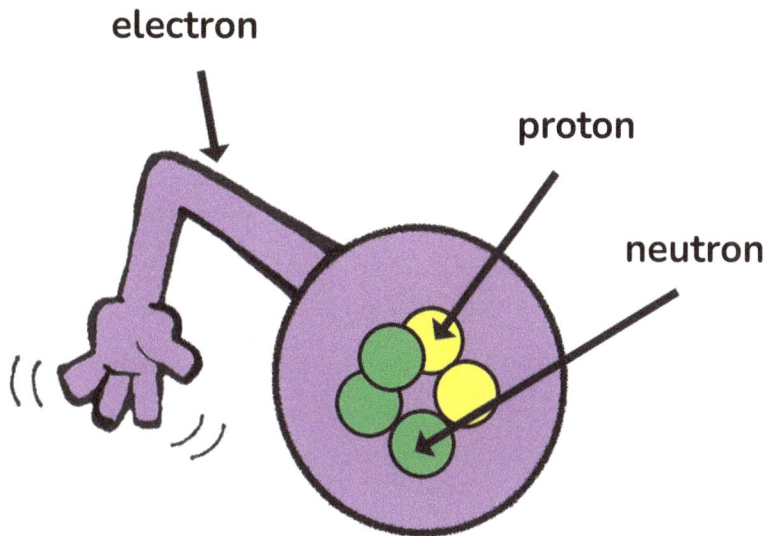

▶ Protons, neutrons, and electrons are the basic parts of atoms. Protons and neutrons are in the center of an atom. Electrons are on the outside.

In our model, the protons and neutrons are shown inside the ball of the atom, and the electrons are shown on the outside as the arms.

The arms in our models represent only the electrons that can help hook an atom to another atom. Some atoms have additional electrons that won't hook to another atom, and these electrons are not shown in our models.

3.4 Atoms Are Different

Atoms are different from each other because they have different numbers of protons, neutrons, and electrons.

one electron

one proton

A model of a hydrogen atom
shown with one "arm" one proton,
and given the color gray

For example, hydrogen only has one proton and one electron. It doesn't have any neutrons.

Carbon has six protons, six neutrons, and six electrons. However, only four of those electrons can help carbon hook to other atoms, so in our model carbon has only four arms.

6 protons

6 neutrons

4 electrons that can form bonds

A model of a carbon atom
shown with four "arms," six protons,
six neutrons and given the color purple

Uranium has 92 protons, 92 electrons, and 146 neutrons! The electrons in uranium can hook to two atoms, four atoms, or even six atoms!

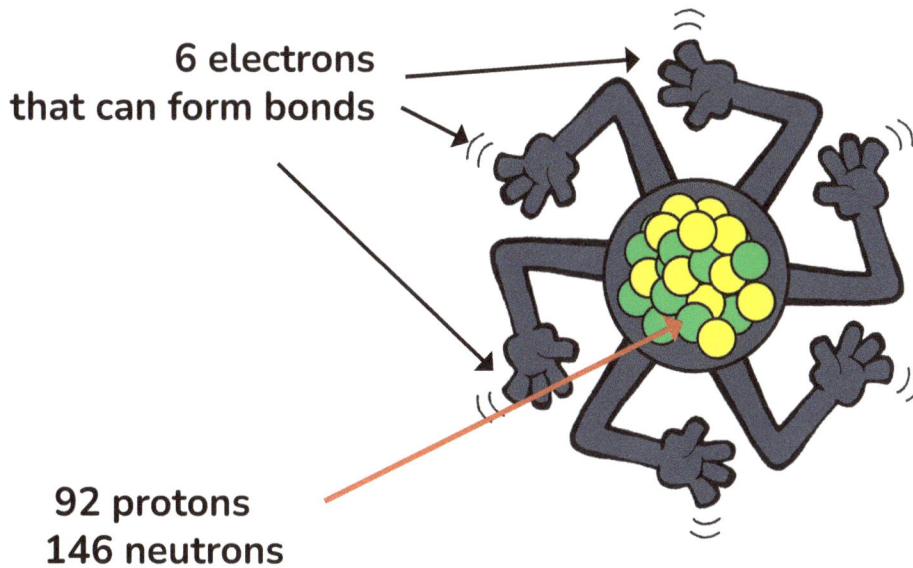

6 electrons
that can form bonds

92 protons
146 neutrons

A model of a uranium atom
with six "arms", 92 protons, 146 neutrons,
and given the color dark green

Because atoms are made of the same basic parts, all atoms are similar to each other. But because atoms are also made of different numbers of those parts, atoms are also different from each other.

Everything you can touch with your fingers, see with your eyes or smell with your nose is made of atoms. All of these things are different from each other because their atoms are arranged in different ways.

3.5 Summary

- Everything we touch, taste, see, or smell is made of **atoms**.

- **Atoms** are very tiny things we can't see with our eyes.

- All **atoms** are made of **protons**, **neutrons**, and **electrons**.

- **Atoms** are different from each other because they have different numbers of **protons**, **neutrons**, and **electrons.**

3.6 Vocabulary

Basic Words

atom (AA-tum)

carbon (KAHR-buhn)

model (MAH-duhl)

proton (PROH-tahn)

Challenge Words

electron (ih-LEHK-trahn)

hydrogen (HIY-druh-juhn)

neutron (NOO-trahn)

nitrogen (NIY-truh-juhn)

oxygen (AHK-sih-juhn)

We are made of atoms!

What Are
MOLECULES?

4.1 Introduction

In Chapter 3 we saw that everything is made of **atoms**. The Moon is made of atoms. Clouds are made of atoms. Carrots, peas, brussels sprouts, and asparagus are all made of atoms.

We also saw that all atoms are made of **protons**, **neutrons**, and **electrons**. We saw that atoms are different from each other when they have different numbers of protons, neutrons, and electrons.

6 protons

6 neutrons

4 electrons that can form bonds

Model of a Carbon Atom

But how do atoms make so many different things?

If we think of atoms as building blocks, we can begin to understand how atoms can make so many different things.

Building blocks are designed to hook to each other to make toy buildings, toy cars, or toy boats. If you look carefully at a building block, you can see that the pegs of one block fit into the holes on another block. By hooking building blocks together, different objects can be created.

In the same way that building blocks hook together to make toy cars or buildings, atoms hook to other atoms to make clouds, carrots, peas, candy canes, and the Moon!

4.2 Atoms Form Molecules

▶ When one atom hooks to one or more other atoms, they form a **molecule**.

▶ Two hydrogen atoms and one oxygen atom hook together to make one water molecule.

Billions of water molecules are in a glass of water, moving around fast and bumping into each other.

▶ **Table salt**, also called sodium chloride, is made of two atoms — **sodium** and **chlorine**.

sodium   chlorine

table salt crystal
(sodium chloride)

Microscope image of table salt crystals

Bigger molecules are made of more atoms.

Glucose, a type of sugar, is made of 24 atoms. It has 6 carbon atoms, 6 oxygen atoms, and 12 hydrogen atoms. (Count them in the illustration below.) Glucose is a type of **carbohydrate**. Carbohydrates are made of carbon and water. Carbohydrates give your body energy. (See Chapter 12).

glucose molecule

Some of the molecules in your body, such as **proteins**, are made of hundreds and thousands of atoms.

A blood clotting protein molecule called anti-hemophilic factor, AHF. Each small ball represents an atom.

4.3 Atoms Follow Rules

▶ Atoms hook to other atoms by following rules!

▶ **Hydrogen** atoms can only hook to **one** other atom. Hydrogen cannot hook to two atoms, three atoms, or more than three atoms.

hydrogen   hydrogen

▶ **Oxygen** can hook to **two** other atoms. But oxygen cannot hook to three atoms, four atoms, or more than four atoms.

oxygen

hydrogen   hydrogen

▶ **Nitrogen** can hook to one, two, or three atoms, but nitrogen cannot hook to more than three atoms.

▶ **Carbon** can hook to one, two, three, or four atoms, but carbon cannot hook to more than four atoms.

Atoms can make many different shapes and sizes of molecules. However, atoms always obey rules when they make molecules. Following the rules means that table salt will always be table salt and sugar will always be sugar!

4.4 Summary

- Atoms hook to other atoms to make **molecules**.

- Two atoms or many atoms can hook together to form a **molecule**.

- Atoms have to **obey rules** when making **molecules**.

4.5 Vocabulary

Basic Words

chlorine   (KLAW-reen)

crystal   (KRIH-stuhl)

glucose   (GLOO-kohs)

protein   (PROH-teen)

silicon   (SIH-lih-kahn)

sodium   (SOH-dee-uhm)

Challenge Words

alcohol   (AAL-kuh-hawl)

carbohydrate
    (kahr-boh-HIY-drayt)

molecule   (MAH-lih-kyool)

Atoms have rules!

Molecules MEET

5.1 When Molecules Meet

In the last chapter we saw that atoms hook together to make molecules. We also found that atoms must obey rules. Each atom hooks to other atoms in its own way.

But what happens when one molecule meets another molecule? What do they do? Do they change, or do they stay the same?

Sometimes when one molecule meets another molecule, they **react**. This means that something changes in the way the atoms are hooked together.

5.2 Molecules Switch Atoms

Sometimes molecules react by switching atoms (exchanging partners). In the example below, two molecules meet and trade atoms. As a result, two **new molecules** are made.

1 Two molecules meet.

hydrogen   chlorine

Hydrogen chloride

sodium

oxygen

hydrogen

Sodium hydroxide

2 The hydrogen atom and the sodium atom switch places.

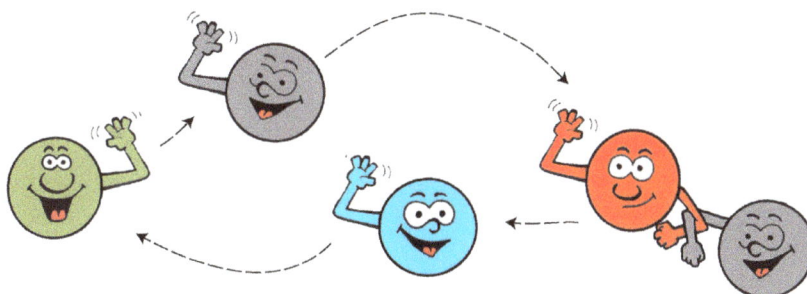

3 Two new molecules are made.

sodium   chlorine

Sodium chloride
(table salt)

Water

hydrogen

oxygen

hydrogen

5.3 Molecules Join Together

Sometimes when molecules meet, they join together. In this example, two chlorine atoms in a chlorine gas molecule meet two sodium atoms. The chlorine atoms and the sodium atoms combine to make table salt.

1 Two chlorine atoms (in a chlorine gas molecule) meet two sodium atoms.

chlorine   chlorine   sodium   sodium

Chlorine gas molecule   **Sodium atoms**

2 The chlorine atoms join the sodium atoms to make two sodium chloride molecules (table salt).

sodium   chlorine

sodium   chlorine

Table salt

5.4 Molecules Break Apart

Sometimes molecules simply break apart to form new molecules. In the illustration below, two water molecules break apart, and then the atoms join together in a different way to make oxygen gas and hydrogen gas.

1 Two water molecules.

2 The water molecules break apart.

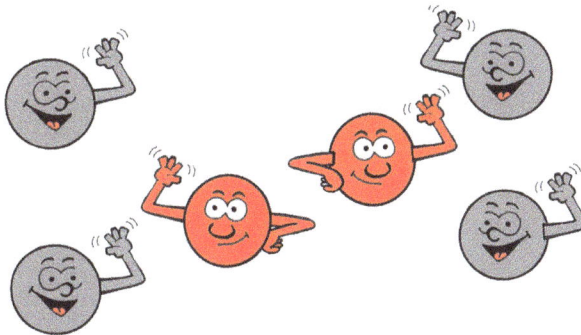

3 New molecules are made.

Hydrogen
gas

Oxygen
gas

All of these examples show how molecules and atoms react with one another.

▶ It is important to realize that in every reaction **atoms are neither created nor destroyed**.

Atoms can rearrange themselves by combining in different ways and changing places to make new molecules, but atoms never simply appear or disappear.

5.5 Reactions Are Everywhere

▶ When atoms join together, leave a molecule, or switch places in molecules, a **chemical reaction** occurs.

There are chemical reactions everywhere. They go on all the time and all around us.

The gasoline inside a car reacts with oxygen to provide energy for the car to move.

Piston ignition time of car engine

Chemical reactions occur when you bake bread or cook an egg or fry a piece of chicken.

Fried chicken

Old rusty shovel in the ground

When you leave your metal shovel out in the rain, the red rust that forms on it is caused by a chemical reaction.

Reactions occur inside your body. When you eat a piece of cheese or drink a glass of milk, reactions occur inside your mouth. These reactions help break down the food molecules into smaller pieces. Inside your stomach there are strong molecules that break your food down still further.

Even when you breathe, reactions inside your lungs help oxygen get into your blood so it can be carried through your body. Reactions are everywhere.

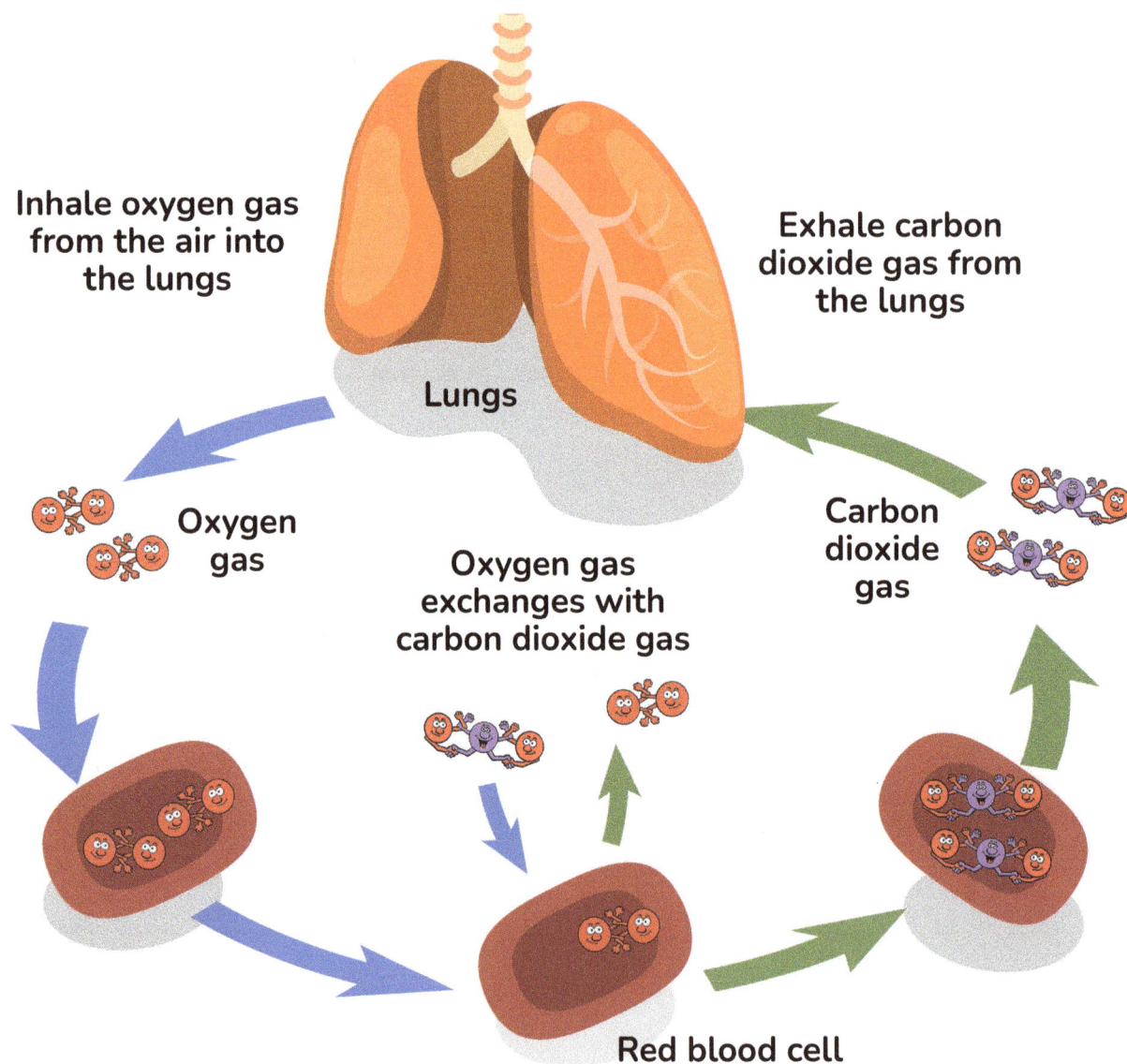

Inhale oxygen gas from the air into the lungs

Exhale carbon dioxide gas from the lungs

Lungs

Oxygen gas

Oxygen gas exchanges with carbon dioxide gas

Carbon dioxide gas

Red blood cell

5.6 Reactions Follow Rules

Reactions also have to follow rules. Not every molecule will react with every other molecule or atom. Some molecules won't react at all. For example, the **noble gases**, such as **neon**, **helium**, and **argon**, usually don't react with any other molecules.

5.7 We Can See Reactions

Often we can observe something happening if a reaction is occurring. Sometimes we can see **bubbles**. Sometimes we might see little **particles** form that look like sand. Sometimes the glass might change **temperature** in our hands. There could also be **fire**, an **explosion**, or a **color change**.

Chemical reaction making smoke

5.8 Summary

- When atoms get rearranged, a **chemical reaction** has occurred.

- Atoms can switch places, join together, or separate from each other during a **chemical reaction**.

- In a **chemical reaction** atoms rearrange but are **never created or destroyed**.

- **Reactions** occur everywhere.

- We may be seeing a **reaction** take place when there are **bubbles**, **color changes**, or **temperature changes**.

5.9 Vocabulary

Basic Words

argon   (AHR-gahn)

neon   (NEE-ahn)

noble gas   (NOH-buhl GAAS)

Challenge Words

explosion   (ihk-SPLOH-zhuhn)

helium   (HEE-lee-uhm)

particle   (PAHR-tih-kuhl)

reaction   (ree-AAK-shuhn)

I love to do science!

What Is BIOLOGY?

6.1 Introduction

In Chapter 1 you learned that science is a way to study how things work, and in Chapters 2-5 you learned the first building block of science — chemistry.

In this chapter we will take a look at the second building block of science called **biology**.

Biology is the study of life. **Biologists** look at how plants and animals live, grow, and interact with each other.

6.2 History of Biology

People have been studying life for thousands of years. The first biologist from Greece was **Aristotle**. Aristotle lived in Greece from 384-322 B.C.E. Aristotle thought about the world around him, and he was particularly interested in living things.

But Aristotle didn't just think about living things, he also made **observations**. Making an observation happens when you examine something with your eyes, or touch something with your fingers, or smell something with your nose.

Aristotle made many observations of both plants and animals. He looked at the parts of animals, how animals move, and how animals breathe or sleep. He also observed how some

Aristotle 384-322 B.C.E.

animals are similar to each other and how some animals are different.

Aristotle had many ideas about life, and he wrote many books to describe his ideas to others. In the same way, modern biologists observe life and write their ideas in books and papers so they can share those ideas with other biologists and other scientists.

6.3 Modern Biology

Today, modern biologists continue to observe life, like Aristotle did thousands of years ago. However, unlike Aristotle, modern biologists can use chemistry and physics to help understand how living things work.

Modern biologists can study how plants make food from light by studying the chemicals in plant leaves and the physics of light waves.

Modern biologists have many tools that Aristotle didn't have. With these tools it is possible for modern biologists to make many more observations about life than Aristotle was able to.

With a microscope modern biologists can see life that is too tiny to see with only our eyes, such as tiny bacteria.

Modern biologists can also observe big animals using tools such as airplanes, boats and submarines.

Close-up of a laboratory microscope and a sample to be studied

Robot inspecting coral deep in the ocean

Biologists can now go to the bottom of the ocean to observe life. Modern biologists use many different tools to study life all over the planet.

6.4 Everyday Biology

Learning about life is something everyone does. Caring for a pet or observing an ant carry away your picnic pie are ways you do biology.

▶ When you watch a frog jump from a branch onto a leaf, you are doing biology.

Red-eyed tree frog

▶ If you make yogurt or watch yeast bubble in a bowl of warm water, you are doing biology.

Yeast bubbling in warm water

You do biology by observing, studying, and working with living things. Biology can involve very complicated experiments or be as simple as observing a moth trapped in a spiderweb. The more observations you make and the more details you observe, the more you will discover about biology.

6.5 Summary

- **Biology** is the study of life.

- **Aristotle** is considered to have been the first **biologist**.

- Modern **biologists** use **chemistry** and **physics** to understand living things.

- **Biology** is observing, studying, and working with living things.

6.6 Vocabulary

Basic Words

physics   (FIH-zihks)

Challenge Words

Aristotle   (EHR-uh-stah-tuhl)

biologist   (biy-AH-luh-jihst)

biology   (biy-AH-luh-jee)

I am a living thing!

Me too!

What Is
LIFE?

7.1 Studying Life

What makes plants, dogs, and beetles different from rocks, dirt, and metal? Maybe you have noticed that rocks don't move like dogs and dirt doesn't need food. Maybe you have seen that forks and knives, which are made of metals, don't crawl around in the kitchen like beetles. **Living things** are different from rocks and dirt and metals because living things are **alive**.

What does it mean to be alive? Think about how you are different from a rock. You need food and a rock doesn't. One feature of being alive is **needing food**.

Second, you can walk, run, jump, curl up into a ball, and roll on the carpet. But a rock can't move. So another feature of being alive is the **ability to move**.

Finally, a rock can't make baby rocks, but plants, animals, and humans all make baby plants or baby animals or baby humans. So another feature of being alive is the **ability to reproduce**.

▶ Living things are different from **nonliving things**.

7.2 Sorting Living Things

How do we keep track of all of the living things we find on the planet? Is there a way to sort them? Why would we want to sort them?

Sorting living things helps us understand how they are different and how they are the same.

For example, what if you had several blocks with different colors, sizes, and shapes. How would you sort them?

▶ You can sort the blocks by **color**.

Red colored blocks   **Blue colored blocks**

▶ You can also sort the blocks by **shape.**

Triangular shaped blocks   **Cube shaped blocks**

▶ By sorting the blocks by shape or color, we can see how some blocks are different (different color), and how some blocks are the same (same shape).

A very long time ago a man named **Carolus Linneaus** thought about how to sort living things. He came up with a system for sorting all of the creatures on the planet. We call this system **taxonomy**.

Taxonomy is a branch of biology that is concerned with how to sort living things.

We sort creatures by looking at different **features**. A feature is anything such as hair, segmented legs, wings, or eyes.

Close-up image of a bee collecting pollen

We sort animals that have hair, like cats, from animals that don't have hair, like snakes. We sort plants that have flowers, like roses, from plants that don't have flowers, like grass. We sort animals that have wings, like bees, from animals that don't have wings, like frogs.

7.3 Domains and Kingdoms

It can be difficult to decide how to sort living things because living things have so many different features. Once upon a time, living things were sorted into only two large groups — **plants** and **animals**. However, as scientists learned more about different living things, they had to make more groups. Today, scientists use three large groups to sort living things. These groups are called **domains**.

▶ The domains are **Eukarya**, **Bacteria**, and **Archaea**.

These domains are further divided into six **kingdoms**. The six kingdoms organize all the known plants, animals, bacteria, fungi, and archaea.

▶ The kingdoms are called **Plantae**, **Animalia**, **Protista**, **Fungi**, **Bacteria**, & **Archaea**.

Houseplants

The kingdom **Plantae** groups all of the plants. Houseplants, trees, cactuses, grass, and ferns are examples of plants in this kingdom.

Animalia groups all of the animals. Dogs, cats, frogs and butterflies, birds and snakes are in the kingdom Animalia.

Blue viper snake

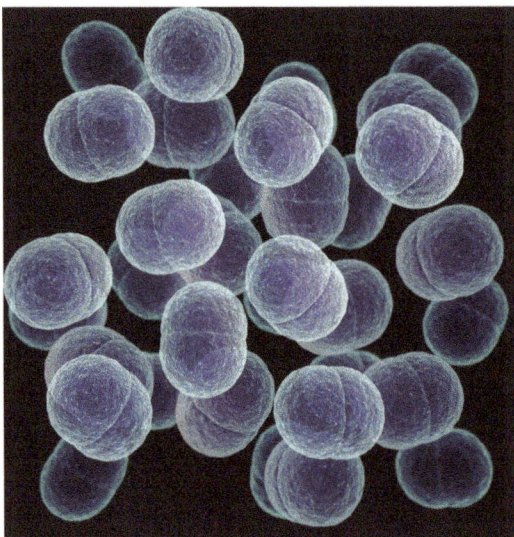

Microscope image of bacteria

Bacteria and **Archaea** are kingdoms that group some of the very small creatures that we can't see when using only our eyes.

Microscope image of an ameoba
(a protist)

Protista is a group for very small creatures called **protists** or **protozoa**. Protozoa live in ponds and streams and can even live on your toothbrush!

The kingdom **Fungi** groups things like toadstools, mushrooms, and yeast. Fungi are found on the forest floor, on dead or decaying trees, and even on your skin.

Red mushroom fungi

7.4 Sorting Within Kingdoms

After living things are sorted into their domains and then into their kingdoms, they are further sorted into smaller groups. To sort living things into smaller groups, scientists again look for different or similar features.

Both birds and cats are animals, but birds are also different from cats. Birds have wings and fly and cats have claws, teeth, fur and eat birds.

Cat and African gray parrot

What about tigers and house cats? They are both cats. Are they exactly the same and in the same group?

Even though tigers and house cats have some similar features, they are also different. House cats don't usually eat their owners, but tigers could! House cats and tigers are put into even smaller groups within the larger grouping of "cats."

7.5 Naming

How do we name all living
things? Because there are
so many different languages
and so many different
living things, scientists
use a **scientific name** for
each living thing. Every
plant and animal, fungus,
bacterium, and archaean
has a scientific name. The
scientific name for each
living thing comes from
Latin. Each living thing has two Latin names. The first name is
called the **genus**, and the second name is called the **species**.

The Latin name for a house cat is *Felis catus*, and the Latin
name for humans is *Homo sapiens*, which means "man wise."

7.6 Summary

- **Living things** are different from **nonliving things**.
  All **living things** need food and can reproduce, and some
  living things can move.

- Scientists sort **living things** into groups to understand them
  better and organize them to study.

- **Domains** and **kingdoms** are two kinds of groups that scientists use to sort living things.

- All living things have a special scientific name.

7.7 Vocabulary

Basic Words

domain   (doh-MAYN)

Fungi   (FUHN-jiy)

genus   (JEE-nuhs)

kingdom   (KIHNG-duhm)

Plantae   (PLAAN-tiy)

species   (SPEE-sheez)

Challenge Words

Animalia   (aa-nuh-MAHL-yuh)

Archaea   (ahr-KEE-uh)

Bacteria   (baak-TIHR-ee-uh)

Eukarya   (yoo-KAA-ree-uh)

protist   (PROH-tihst)

Protista   (proh-TEE-stuh)

protozoa   (proh-tuh-ZOH-uh)

taxonomy   (taak-SAH-nuh-mee)

What's your kingdom?

If I tell you, will you put me down?

The Cell
A TINY CITY

8.1 Living Things

Have you ever wondered what makes a frog a frog or a rose a rose? Have you ever wondered what you are made of?

What are living things made of?

▶   Living things are made of **atoms.**

We know from chemistry that atoms fit together to make molecules. In this chapter we will explore how molecules fit together to make living things like frogs.

8.2 The Cell

The atoms and molecules in living things are designed to fit together to make **cells**.

Atoms and molecules inside cells are very small and can't be seen. Together atoms and molecules make cells. Cells fit together to make all the parts of living things.

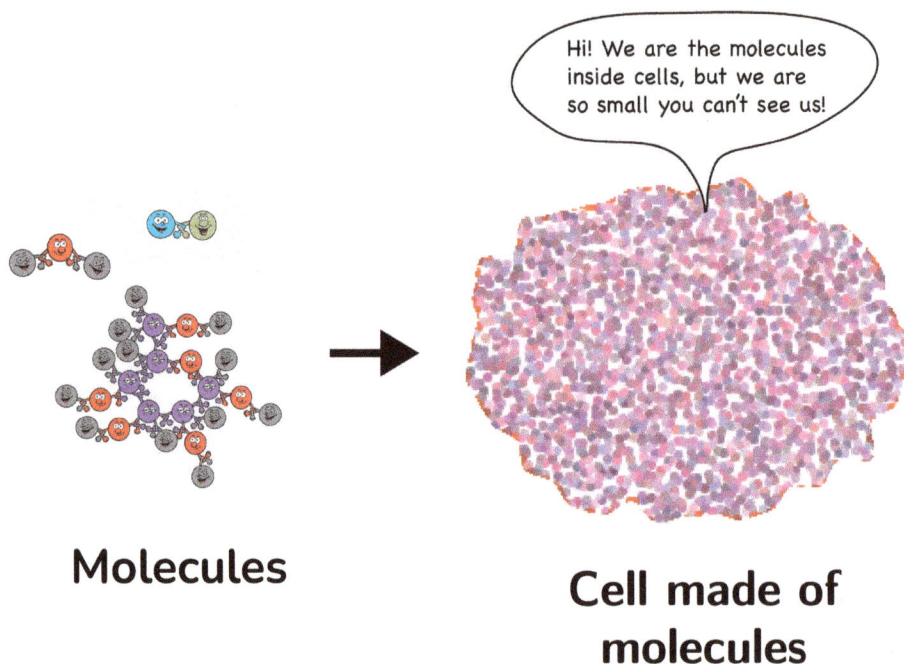

Hi! We are the molecules inside cells, but we are so small you can't see us!

Molecules

Cell made of molecules

▶ Molecules fit together to make skin cells and skin cells fit together to make skin.

▶ Molecules also fit together to make muscle cells and muscle cells fit together to make muscles.

▶ Bone cells, muscle cells, and skin cells fit together to make arms or legs or fingers.

Bone cell

Muscle cells

Skin cells

8.3 A Tiny City

Cells are like tiny cities. Imagine what it takes to make even one item we use in a city — like milk. What does it take to make milk?

We know that milk comes from cows.

Cows usually live in the country, so we have to go to the country to get the milk from the cows.

Once we have the milk, we have to move it to the city. We need a big truck to move the milk.

It can't be any kind of truck, it has to be a special dairy truck. If it were a sand truck, we might get sand in our milk. We have to use a milk truck.

Once the milk is in the city, what happens to it? It would be hard to get our milk directly from the milk truck, so the milk is put into cartons.

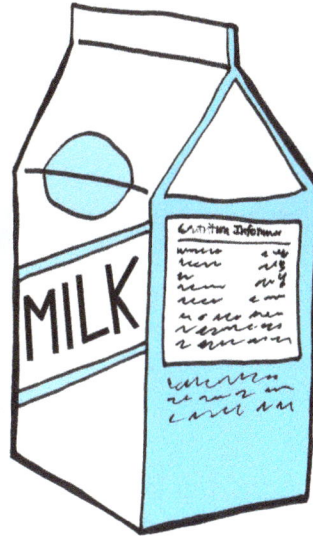

When the milk is put into cartons, we have to put it somewhere we can find it, like a grocery store. When it is in the grocery store, we can buy it and put it on our table and drink it.

It takes many steps to bring milk to your table. Someone needs to milk the cows. Someone needs to clean the raw milk and move it to a milk truck. Someone needs to drive the truck into town, put the milk into cartons, and put the milk in the grocery store.

It takes a lot of people to perform all those jobs! All the jobs have to all be done in the right order. Imagine what would happen if the milk was put into cartons in the wrong order. What would happen if the cartons were sealed before the milk was put in?

The process for making milk is similar to how a cell works. There are lots of complicated molecules called **proteins** doing all kinds of jobs inside a cell that keep the cell alive.

Each job has to be done by a special protein, and all the jobs have to be done in a particular way and in a particular order.

8.4 Parts of a Cell

Cells have different parts like cities have different places. We have talked about how cities provide milk in a grocery store, but there are also other places in cities that provide other things, like shoes, basketballs, and cereal.

In cities, there is usually a downtown area with many important places, like the courthouse or the tax department. Scattered through the city are post offices where letters are mailed back and forth, grocery stores for buying milk, hotels for people who visit, and factories for making things.

▶ In a city there are different places and different people doing different jobs. *Cells are like cities.*

▶ In a cell there are different places with different molecules doing different jobs.

In most cells a central part does many of the important jobs in the cell. This central part is called the **nucleus**. The nucleus of human and plant cells holds all of the important information, called **DNA**, for the cell. The nucleus is like a DNA factory.

Nucleus (building)

Golgi apparatus (building)
Ribosomes (protein man)

Outside the nucleus there are lots of places that do other special jobs. There is a protein factory that includes the **Golgi apparatus** which makes proteins, and **ribosomes** that make other proteins.

Kinesin (truck)
Microtubules (road)

There are "roads," called **microtubules**, that move proteins from place to place. And there are "trucks" called **kinesin** that carry the proteins on the "roads."

Putting it all together, the cell looks a lot like a little city with every protein and molecule doing their part!

A City Cell

8.5 Summary

- Atoms and molecules fit together to make **cells**.

- Cells fit together in special ways to make the parts for living things.

- Each part of a cell has a special job.

8.6 Vocabulary

Basic Words

cell   (SEHL)

protein   (PROH-teen)

Challenge Words

DNA   (DEE-EHN-AY)

Golgi apparatus   (GOHL-jee aa-puh-RAA-tuhs)

kinesin   (kih-NEE-suhn)

microtubule   (miy-kroh-TOO-byool)

nucleus   (NOO-klee-uhs)

ribosome   (RIY-buh-sohm)

Awesome new biology words!

Viruses, Bacteria & Archaea

9.1 Introduction

You might think the smallest form of life is a mouse or a plant. But there are smaller forms of life. You can't see these small types of life with only your eyes. What are they?

Have you ever wondered what happens when you catch a cold or get an infection after you get a cut? Colds and infections are caused by two of the smallest forms of life — **viruses** and **bacteria**.

Archaea are another tiny form of life. Viruses, bacteria, and archaea can only be seen with a **microscope**. A microscope is a tool that makes tiny things look bigger so we can see them.

9.2 Viruses

Viruses are simple and small. Viruses don't have all the parts of a complete cell. Viruses are a bag of molecules covered with more molecules. Because viruses are not true cells, they must live inside other **organisms** (living things) to survive.

Scientists argue about whether or not viruses can really be called "alive." Viruses are usually classified according to the type of organism they are living inside, and viruses don't have their own kingdom.

▶ Viruses can cause diseases in plants, animals, and people.

The common cold virus is small and looks like a soccer ball with spikes. This virus can live for many days on nonliving objects, like door handles or pencils.

When you touch an object that has a cold virus on it, the virus can enter your body, especially if you touch your mouth or face.

If the virus has time to settle in, it will start to multiply in your body and make you sick. Coughing and sneezing are ways that your body uses to try to get rid of the virus.

▶ The best way to keep from catching a cold virus is to wash your hands often!

9.3 Bacteria

Bacteria are larger than viruses. Bacteria have all the parts of true cells and are considered by all scientists to be alive. Bacteria eat, build molecules, grow, and divide to make new bacteria. Some bacteria can even move on their own.

Bacteria have different shapes. Some are rod-shaped, like hot dogs. Some are round, like balls. Some are spiral shaped, like a spring.

Some bacteria stay by themselves. Other bacteria like to stick to each other to form colonies. A **colony** of bacteria is just a large group of individual bacteria that are all living very close together in a group.

Bacteria live everywhere. Some bacteria live in the dirt and help plants get food from the soil. Without bacteria, plants could not get the food they need to grow.

Bacteria also live in water. Bacteria in water are sometimes food for other living things. Sometimes bacteria in water can make us sick. Cities clean our drinking water to remove bacteria that could make us sick.

Bacteria live on us and inside of us. Bacteria live on our skin, in our mouths, throats, stomachs, and intestines. Some bacteria help us digest our food.

When bacteria live under the right conditions, they will divide and make more bacteria.

To stay alive, all bacteria need to have water available. Some bacteria can survive when there is a lack of water, but they will not grow and divide until more water becomes available.

9.4 Archaea

Archaea are similar to bacteria but are different enough to be classified in a group of their own.

Archaea have cell types that are slightly different from those of bacteria. Archaea are often found in places that have very extreme conditions, like places that are very hot or very salty. Some archaea even eat sulfur or iron for food!

9.5 Summary

- **Viruses**, **bacteria**, and **archaea** are tiny **organisms**.

- **Viruses**, **bacteria**, and **archaea** are able to live under many different conditions.

- **Viruses** typically cause illness.

- Some **bacteria** are helpful. Other **bacteria** can cause illness.

- **Archaea** are both similar to and different from **bacteria**.

9.6 Vocabulary

Basic Words

virus   (VIY-ruhs)

Challenge Words

bacteria   (baak-TIHR-ee-uh)

colony   (KAH-luh-nee)

microscope   (MIY-kruh-skohp)

organism   (AWR-guh-nih-zuhm)

I am glad there are lots of good bacteria!

What Is PHYSICS?

10.1 Introduction

In the previous chapters we explored the first two building blocks of science — chemistry and biology. In this chapter we will take a look at the third building block of science: **physics**.

Have you ever thrown a ball up in the air? Did you notice the ball when it left your hand? What did it do? Did it go up? Did it come back down? Unless it gets stuck in a tree or picked up by a big bird, a ball that is thrown up into the air will always come back down.

Have you ever tried to throw a ball really far or really high? Have you ever watched how far or how high the ball goes? Have you ever noticed that it's harder to throw a heavy ball than it is to throw a light ball? Have you ever noticed that it's almost impossible to throw a feather?

Physics is the branch of science that explores how far or how high a ball might go or how heavy it needs to be so that it can be thrown. Scientists who study physics are called **physicists**.

10.2 History of Physics

Physics is about studying the way things behave and then figuring out the rules those things follow to make them behave that way. Physicists don't make the rules, but they discover the rules by watching how the world works.

Aristotle studied motion, but it was **Galileo Galilei,** an Italian astronomer, who used physics to understand how things move.

Galileo is known for a famous experiment where he dropped two balls off a building to see what would happen. He used a heavy ball and a light ball. To everyone's surprise, he found out that they both hit the ground at the same time!

Physicists also use math to figure out the rules. **Isaac Newton** was a great scientist and mathematician who figured out many important rules of physics. By using math, Newton figured out exactly why the balls Galileo dropped hit the ground at the same time.

▶ Math is an essential part of physics and helps us understand the rules of physics.

10.3 Modern Physics

Did you know that balls will follow the same rules no matter where you are on the Earth? You can be in the frozen Arctic, and if you drop two balls, they will fall in exactly the same way. You can be in a desert, at the beach, or on a boat, and if you drop two balls, they will fall in exactly the same way. No matter where you are on Earth, a ball will always follow the rules of physics!

▶ How balls behave when dropped is explained by the rule "what goes up - comes down" which is a rule about gravity. **Gravity** is what makes the balls come back down.

Gravity is also what keeps you from flying off the surface of the Earth. Balls, toys, cars, houses, and even birds obey the rules of gravity.

10.4 Everyday Physics

Every day, physics is happening all around you. In the same way that you learn chemistry and biology, to learn physics you need to make **observations**. When you are looking at something with your eyes, you are making an observation. Making good observations is the first and most important step when you are trying to understand physics.

When you make an observation, try to notice everything you can about what you are observing. If you are at the movie theater getting popcorn, try noticing the popcorn machine. Where does the popcorn go in? Where does it come out? What is moving on the machine? What is staying still? Notice the popcorn when it comes out. Is it hot or is it cold? These kinds of observations are important if you want to think like a scientist.

10.5 Summary

- **Physics** is about studying the way things behave and then figuring out the rules those things follow that make them behave that way.

- Objects, like balls, planes, and birds, always obey the **rules of physics.**

- The **rules of physics** are true no matter where you are on the Earth.

- **Physicists** don't make the rules, they discover the rules.

- To think like a scientist, you must make good **observations**.

10.6 Vocabulary

Basic Words

physics   (FIH-zihks)

Challenge Words

Aristotle   (EHR-uh-stah-tuhl)

Galileo Galilei   (gaa-luh-LAY-oh gaa-luh-LAY)

gravity   (GRAA-vuh-tee)

Isaac Newton   (IY-zuhk NOO-tuhn)

physicist   (FIH-zuh-sihst)

I love to learn physics!

Push and PULL

11.1 Up the Hill

What do you think would happen if you tried to take your baby sister up a hill in a wagon? You might start on flat ground at the bottom of the hill. First, you might grab the handle of the wagon, and as you did so, you might feel the wagon pull against you. You might think that the wagon with your sister in it is heavy.

Then, to get the wagon moving by pulling on it, you might need to use all of the strength in your arms and legs.

Once the wagon is moving on flat ground, you might find that it is easy to roll your sister along and that you don't have to use as many muscles. But, as you get to the bottom of the hill, you might need to use all your muscles again to pull the wagon uphill.

When you reach the top of the hill, you might discover that you are completely out of breath and a little tired. You might say that you used all your energy to do the work you needed to do to get your sister up the hill. And a physicist would say you are exactly right!

When you pull your little sister up a hill in a wagon, you are doing work and you are using energy.

▶ In physics, **work** is what happens when **force** moves an object.

▶ **Energy** gives you the ability to do work.

But what is force? And what is energy?

11.2 Force

When you are pulling the wagon up the hill, you are using a force. **Force** is any action that changes the location of an object. Because you are changing the location of the wagon (and your sister) you are using force.

Force is also any action that changes the shape of an object. If you were to squeeze a marshmallow, you would be using force. By squeezing a marshmallow, you are changing its shape with the force created by your hands.

Force is also any action that changes how fast or how slowly an object is moving. You may have experienced this kind of force if you ever tried to catch the same ball as your friend. If you were both looking at the ball and not where you were going, you might have run into each other.

When that happens – WHAM! – the two of you collide, and you both stop moving. In this action you each used force to stop the other from moving. You could probably feel the effects of the force stay with your head or knees for several hours!

▶ **Force** is any action that changes the location of an object, the shape of an object, or how fast or slowly an object is moving.

11.3 Work

In the first section we saw that **work** happens when force moves an object. **Work** also happens when force changes the shape of an object or when force changes how fast or how slowly an object is moving.

How is work related to force? If you use more force, are you doing more work? Maybe. Or if you are doing more work, are you using more force? Yes!

Imagine that instead of one little sister, you have two. And imagine trying to pull both little sisters up the hill in the wagon.

It will take more muscles, more energy, and more effort to move two little sisters up the hill. In fact, if your little sisters were twins and weighed exactly the same amount, you would have to do twice the amount of work to get both of them up the hill.

The same is true for squeezing a marshmallow or running into your friend. If you squeeze the marshmallow more, you are doing more work. If you and your friend are running faster and generating more force, you are doing more work when you run into each other.

11.4 Energy

How do you get the
energy for pulling your
little sister up the hill in
a wagon or for squeezing
a marshmallow or for
colliding with your friend?
Where does the energy
come from?

The energy for you to do
all these things comes
from your breakfast. When you eat breakfast, you are giving
your body the energy it needs to do work.

▶ In physics, **energy** is something that gives something else
the ability to do work.

When you eat eggs or toast or cereal, your body takes the
energy in the food and uses it in a way that helps your muscles
move.

The food you eat for breakfast gives your muscles the ability
to do work—like pulling a wagon full of sisters up a hill! And
that's a lot of work.

11.5 Summary

- A **force** is any action that changes the location of an object or the shape of an object or how fast or slowly an object is moving.

- **Work** is what happens when a **force** moves an object.

- **Energy** is something that gives something else the ability to do **work**.

11.6 Vocabulary

Basic Words

force　(FAWRSS)

work　(WEHRK)

Challenge Words

energy　(EHN-uhr-jee)

Playing basketball uses force, work, and energy.

Types of ENERGY

12.1 Stored Energy

Recall from the last chapter that energy is something that gives something else the ability to do work. We saw how the food you eat for breakfast gives you the ability to move your muscles, which allows you to pull your little sister up a hill.

The food you eat for breakfast is a type of stored energy.

▶ **Stored energy** is energy that has not been used.

A box of cereal may not look like it is full of energy, but in fact it has lots of energy molecules.

These energy molecules are called **carbohydrates**. When your body needs energy, it will break down the carbohydrate molecules so they can be used for moving muscles or walking to the store.

12.2 Types of Stored Energy

There are different types of stored energy. Have you ever taken a rubber band and stretched it across your finger and thumb? What happens when you release your thumb, letting the rubber band go? It flies through the air. There is stored energy in a rubber band.

But can you use this stored energy for breakfast? NO! You wouldn't eat rubber bands for breakfast!

A rubber band has a different type of stored energy than your breakfast cereal.

Breakfast cereal has chemical stored energy. **Chemical stored energy** is energy that comes from chemicals and chemical reactions.

A rubber band has elastic stored energy. **Elastic stored energy** is energy that is found in materials that can stretch.

A book on a table has gravitational stored energy. It could do work if it were to fall to the ground and smash an egg. **Gravitational stored energy** comes from objects that are elevated above the ground and can be pulled down by gravity.

12.3 Releasing Stored Energy

When you let a rubber band go from your thumb, you release the stored elastic energy and the rubber band flew through the air. When you eat cereal for breakfast, your body breaks down the carbohydrate molecules, releasing the stored chemical energy so your muscles can pull a wagon. If a book falls off a table and onto the floor, the stored gravitational energy of the book is released and can be used to break an egg.

When each of these types of stored energy is released, the energy does not disappear but is converted into moving energy. The rubber band moved, the wagon moved, and the book moved.

In order for stored energy to do work, it must first be released. The stored energy in the rubber band, the breakfast cereal, and the book were all released and changed into moving energy. But what is moving energy?

12.4 Moving Energy

Moving energy is the energy found in moving objects.

▶ There is only one type of moving energy, and physicists call this energy **kinetic energy**.

Kinetic energy can come from different types of stored energy, but only the energy of a moving object is called kinetic energy.

When you release the stored energy in a rubber band and it's sitting on the floor, it can no longer do any work unless it is picked up and stretched again. When the rubber band is sitting on the floor, all of the stored energy has been released, and all of the kinetic energy has been used. You might think that the energy is lost. However, none of the energy disappears, it just gets changed to a different type of energy.

12.5 Summary

- **Stored energy** is energy that has not been used yet.

- There are different types of **stored energy**. Breakfast cereal has **chemical stored energy**. A rubber band has **elastic stored energy**.

- When **stored energy** is released, it can be changed into **kinetic energy** — the energy of an object that is moving.

12.6 Vocabulary

Basic Words

stored   (STAWRD)

Challenge Words

carbohydrate   (kahr-boh-HIY-drayt)

elastic   (ih-LAAS-tihk)

gravitational   (graa-vuh-TAY-shuh-nuhl)

kinetic   (kih-NEH-tihk)

Playing tennis uses kinetic energy.

Saving ENERGY

13.1 Energy to Energy

In the last few chapters you learned how energy is used when a force does work. You also learned about different kinds of energy. You learned about chemical stored energy, elastic stored energy, and gravitational stored energy. You also learned about kinetic, or moving, energy.

You learned that when a stretched rubber band is released, it can no longer do any work. But the energy is not lost, it has just changed to a different type of energy.

For example, if you look carefully at your brother when he is riding a bicycle, you can observe different types of energy being used to generate different kinds of forces and work.

Your brother's body is using **chemical energy** from his breakfast to move his muscles. The muscles are using the chemical energy to move the pedals on the bike. The pedals on the bike are connected to a chain and a gear. As the chain and the gear move, the wheels move. The chain, gear, and wheels are all using **mechanical energy** to move the bike forward.

▶ In this example, you can see how one type of energy is getting **converted**, or changed, into another type of energy.

As we saw in Chapter 12, energy is only converted (or changed) into other types of energy.

▶ Energy is never created, and energy is never destroyed. It is only changed from one form to another.

13.2 Energy We Use

When your dad puts gas in the car, he is giving the car energy it can use. Your body can't use gasoline to move the pedals on a bicycle, but a car uses gasoline to run the motor. Likewise, a car cannot use cereal to move its motor like you use cereal to give energy to your legs.

▶ There are different forms of energy, and not all types of energy can be used in the same way.

Gasoline is one form of energy that is used to power things like cars and boats. Cereal and bread are another form of energy used to power things like human bodies.

Batteries are yet another form of energy used to power things like flashlights and laptop computers.

▶ When we "use" energy, we are converting one form of energy to another form of energy.

When you use batteries in electronic toys, the chemical energy in the batteries is converted to moving and sound energy.

Eventually the batteries run out. All the chemical energy is converted to moving and sound energy, and there is no more chemical energy in the batteries. This is how we "use" energy. We convert it from one form to another form.

13.3 Energy We Waste

Even though energy cannot be destroyed, it is possible to "waste" energy. You might have heard your dad telling you to turn off the lights after you leave a room. Or you might hear your mom tell you not to leave the door open in the middle of winter. They may have told you not to "waste" energy. But if energy cannot be destroyed, what does it mean to waste energy?

▶ Energy is "wasted" when energy is excessively or unnecessarily converted from one form to another form.

If you are playing with your battery powered car, then you are converting one form of energy (chemical energy in the battery) to another form of energy (moving energy in the car). Because you are playing with the car, it is necessary to convert the chemical energy.

But if you walk away from the car, and you forget to turn it off, the battery is running but you are no longer playing with the car. Now you are "wasting" the energy in the battery because you are not using it to play with the car. When you go back to your car the next day, you discover it won't work anymore because the battery is "dead." All the chemical energy in the battery is gone. This is wasting energy. You didn't destroy the energy, you just unnecessarily converted it from chemical to moving energy when you didn't need to.

13.4 Finding Energy

We get much of our energy from the Earth. Gasoline, **coal**, and **natural gas** all come from inside Earth. The fuel for **nuclear energy**, such as **uranium**, comes from the Earth. In reactors, some uranium is transformed into **plutonium**, which can also be used as fuel. It is possible to get energy from moving water or wind, and we can also get energy from the Sun.

The food we eat comes from the Earth. We grow plants and raise animals to get food for our bodies. The plants get their energy from the Earth and also from the Sun.

Will we ever run out of energy? Yes and no. The energy we get from gasoline comes from **fossils**. There are only so many fossils in the ground, so it is possible that one day we will have used all the energy that is stored in fossils. When that happens, we will not have any more gasoline that comes from fossils to run our cars. This is also true with other forms of energy that come from fossils, such as coal or natural gas.

But if we remember that energy cannot be destroyed, only converted from one form to another, it may be possible to discover new ways to convert energy. Maybe there are new ways to convert the Sun's energy to chemical energy. Or maybe there are ways to get chemical energy from rocks. Or maybe there are ways to get electrical energy from grass or trees. Maybe you will be the next scientist who discovers a new way to convert energy into a form that can be used to power cars or boats or planes!

I WONDER IF I CAN CONVERT THESE CHERRY POPS INTO GASOLINE?

13.5 Summary

- Energy is neither created nor destroyed.

- Energy is **converted** from one form to another form.

- Energy is wasted by converting it from one form to another form excessively or unnecessarily.

- There may be new ways to convert energy to a form that can be used for fuel.

13.6 Vocabulary

Basic Words

convert   (kuhn-VUHRT)

fossil   (FAH-suhl)

Challenge Words

mechanical   (mih-KAA-nih-kuhl)

natural   (NAA-chuh-ruhl)

nuclear   (NOO-klee-uhr)

plutonium   (ploo-TOH-nee-uhm)

I have lots of energy!

What Is GEOLOGY?

14.1 Introduction

We have explored three important building blocks for science — chemistry, biology, and physics.

We are now ready to take a look at a fourth building block of science: **geology**.

Geology is the study of the Earth. By studying Earth, scientists attempt to understand what the Earth is made of, how Earth came into being, how Earth has changed in the past, how it is changing now, and our role as we live on Earth.

Do you ever pick up rocks and wonder how they were made and what they are made of? Do you sometimes look at mountains and wonder how they were formed? Have you wondered what's at the bottom of the ocean? Have you noticed how weather affects the landscape? Do you wonder why certain birds and wild animals live near you and others don't? These are all

WOW! THERE ARE **MOUNTAINS** AT THE BOTTOM OF THE OCEAN!

questions that are explored by scientists who study geology.

14.2 History of Geology

Depending on where you live, when you go outside and walk around, you'll see many different features of Earth. You might see mountains or rivers. You might see fields of dirt or fields of grass. You might see lakes or oceans, mesas or glaciers, forests or prairies.

Ancient people also saw many of the same features you see. Although lakes come and go and rivers might change course, many of the features you see today are the same features ancient people would have seen.

One of the first people to study Earth's features was the Greek philosopher **Theophrastus** who lived from about 371-287 BCE.

Theophrastus was a student of **Aristotle**. Recall from Chapter 6 that Aristotle was a Greek philosopher who was one of the first to study plants and animals. Like Aristotle, Theophrastus was interested in science. He studied rocks and explored what happens when rocks are heated.

Theophrastus
371-287 B.C.E.

Many of the first geologists also asked questions about how Earth came into being and how many years Earth has existed. All of these questions shaped the modern science we now call geology.

14.3 Modern Geology

Modern **geologists** continue to study rocks and what rocks are made of. They also ask questions about how mountains, rivers, and glaciers form. Modern geologists have an advantage over ancient geologists because modern geologists can use chemistry and physics to better understand how things work.

I WONDER WHAT THESE ROCKS ARE MADE OF...

There are different kinds of modern geologists. Some modern geologists focus on the chemistry of Earth. These geologists are called **geochemists**. Geochemists study how atoms and molecules make rocks, soils, minerals, and fuels.

Other modern geologists focus on the structure of Earth. These geologists are called **structural geologists**. Structural geologists study how Earth is put together and how it changes. They are interested in how rocks change and what makes mountains and valleys.

I WONDER HOW THIS MOUNTAIN FORMED...

There are also modern geologists who study how humans affect the water, air, and land quality of Earth. These geologists study Earth's environment and are called **environmental geologists**.

I WONDER HOW THIS IRON MINE AFFECTS THE ENVIRONMENT?

OH, GOOD! IRON ORE FOR DINNER!

FERRO

14.4 Everyday Geology

Even though you may not be a geologist yet, you can learn about the Earth by simply observing what happens around you.

What happens when it rains? Do the roads flood? Do you get mud slides, or does a river find a new path? What happens in the hot sun? Do you observe mud forming cracks or rocks crumbling? Have you ever been in an earthquake? Did you feel the ground move?

Paying attention to where you live, what happens during storms, and how the land around you changes over time are activities you can do every day.

14.5 Summary

- **Geology** is the study of Earth.

- The first **geologists** looked at rocks and minerals and asked questions about how Earth came into being.

- **Geochemists** are modern **geologists** who study how atoms and molecules form Earth.

- **Structural geologists** look at how Earth is put together.

- **Environmental geologists** look at changes in the quality of the water, the air, and the land on Earth.

14.6 Vocabulary

Challenge Words

environmental   (ihn-viy-ruhn-MEHN-tuhl)

geochemist   (jee-oh-KEH-mihst)

geologist   (jee-AH-luh-jihst)

geology   (jee-AH-luh-jee)

structural   (STRUHK-chuh-ruhl)

Theophrastus   (thee-uh-FRAA-stuhs)

I love to learn geology words.

What Is Earth Made Of?

15.1 Introduction

If you walk outside, do you notice where you live? If you live in the country, do you notice the trees and grass? If you live in a city, do you notice the buildings and streets?

If you look down, do you notice the ground? If you live in a city, do you notice how much of the ground is covered up with streets or pavement? If you live in the country, do you notice if the ground has grass, rocks, or dirt?

When you notice the ground with the rocks, dirt, grass, and trees, you are noticing the Earth. Earth is where you live.

But what is Earth made of? What is dirt? What are rocks? Why do grass and trees grow in dirt? How deep does the dirt go? How many kinds of rocks are there? What is below the rocks and dirt? More rocks? Trees? Chocolate syrup?

In this chapter we will learn about what Earth is made of.

15.2 Rocks and Minerals

The **crust** is the outer part of Earth and is where we live. The crust is mostly made of rock. If you go outside and start digging in the ground with a shovel, you will eventually hit some type of rock.

What is the difference between a rock and a tree? Trees are living things. Rocks are not living things. Rocks do not move like living things. Rocks do not grow like living things, and rocks do not multiply like living things.

Living things are made mostly of **carbon**, but rocks are made mostly of **silicon**. Silicon and carbon are **atoms** (also called **elements**). Because rocks are made mostly from silicon and living things are made mostly from carbon, rocks are very different from living things.

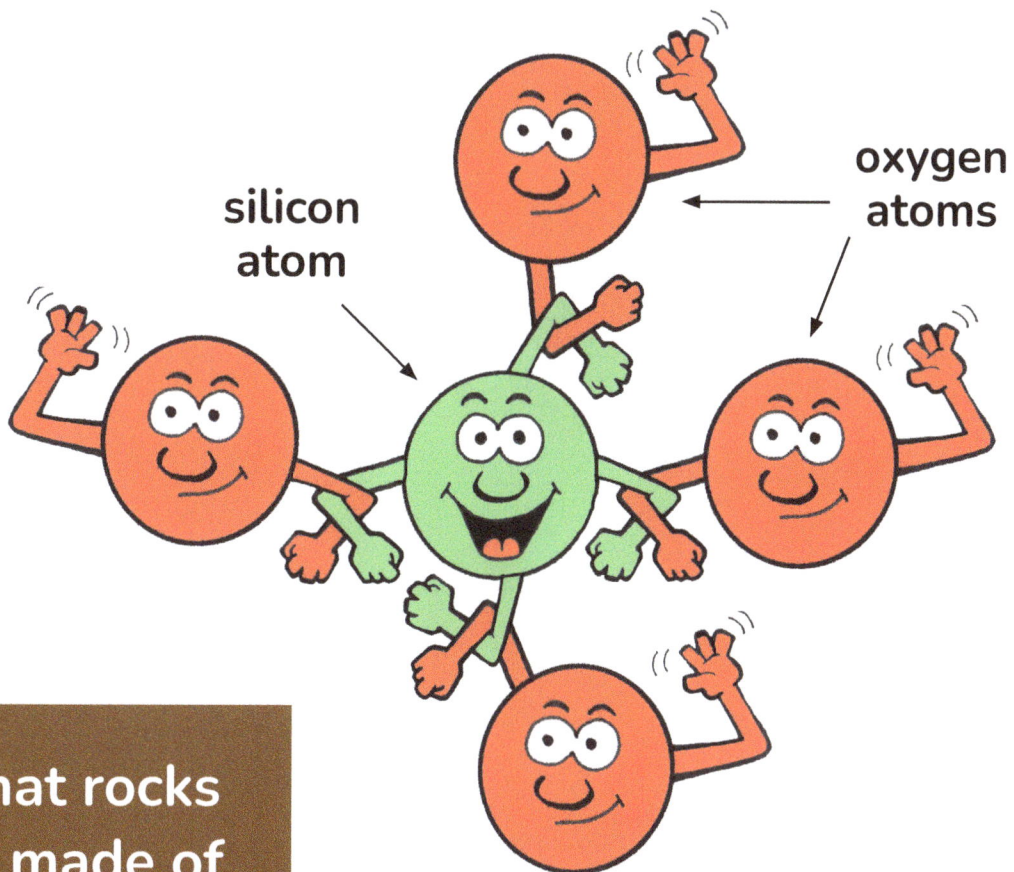

silicon atom

oxygen atoms

What rocks are made of

▶ All rocks come from **magma**, which is molten (melted) rock deep inside the Earth.

Magma is made mostly of the elements silicon and oxygen. Rocks form when the magma cools and mixes with other elements, like magnesium, iron, or aluminum.

When magma cools very slowly, the atoms in the magma have a chance to line up in an orderly fashion. When this happens, the material that is formed is called a mineral.

HOT!!!
HOT!!!
HOT!!!

MAGMA

INSIDE EARTH

There are many different kinds of minerals. One common mineral found in rocks is **quartz**. Quartz can be clear, pink, purple, or other colors.

Mica is also a mineral. Mica is soft and looks like thin, layered paper. You can peel mica sheets away from each other.

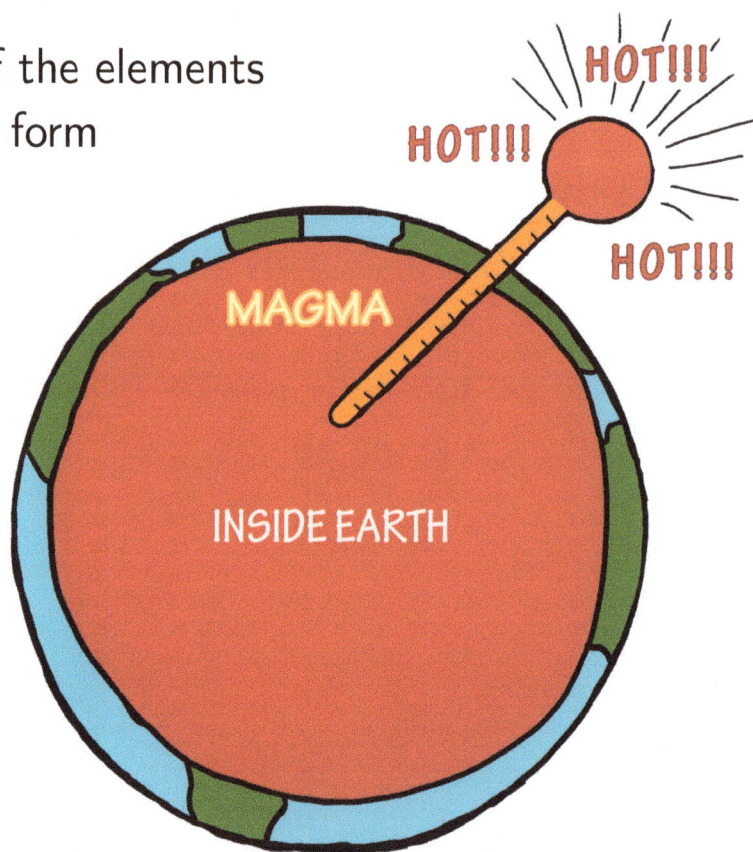

Calcite is another mineral found in rocks. Calcite is made of calcium and oxygen and forms beautiful crystals that come in different colors.

Minerals

Quartz   Mica   Calcite

Some minerals are used in jewelry. **Rubies** are a brilliant red-colored mineral made of **aluminum** and oxygen. **Emeralds** are a different type of mineral made of aluminum, **beryllium**, silicon, and oxygen. Emeralds have a deep green color. Both rubies and emeralds are hard to find, and that is why jewelry made with them is often expensive.

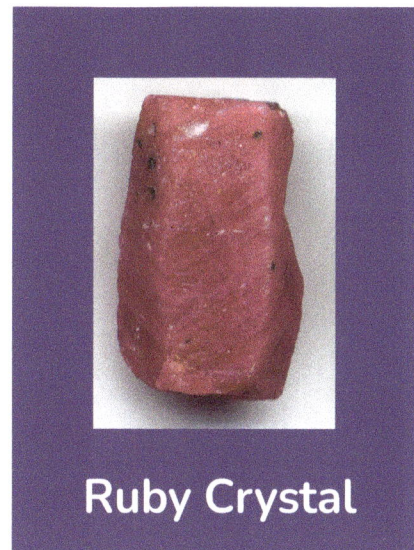

Ruby Crystal

There are also many different kinds of rocks. There are rocks that are formed deep inside the Earth, and there are rocks that form on the surface of the Earth. There are also rocks that have been changed from one type of rock into another type of rock. The different types of rocks are called **igneous, sedimentary**, and **metamorphic**. Minerals are the building blocks of all these types of rocks.

Types of Rock

Igneous
(granite)

Sedimentary
(limestone)

Metamorphic
(garnet schist)

▶ **Igneous** rocks are the most plentiful type of rock.

Igneous rocks are formed when molten magma inside the Earth cools and hardens. Granite is one kind of igneous rock, and it has a lot of quartz in it.

A Sedimentary Rock

▶ **Sedimentary** rocks are formed from bits of rocks and other materials left behind by wind or water.

As these materials pile on top of each other, layers form. These layers are pressed tightly together, turning the materials into rock.

▶ **Metamorphic** rocks are those that have changed from one type of rock into another.

Very high heat and pressure cause these changes. Igneous rocks and sedimentary rocks can be changed into metamorphic rocks. Even metamorphic rocks can be changed into other metamorphic rocks!

A Metamorphic Rock

15.3 Dirt

When you go outside and dig a hole in the ground, you not only find rocks, but you also find dirt. Dirt is made of rocks, plants, minerals, and even dead animals!

There are many different types of dirt. Some dirt is sandy and light in color. Other dirt is moist and dark in color. Have you ever wondered why some parts of the world are used for growing food and other parts of the world are not? That's because some dirt is good for growing plants and other dirt is not.

▶ Dirt is also called **soil**.

Dirt and soil come mainly from rocks. Because different parts of the Earth have different kinds of rocks, there are different types of soil in different places.

15.4 Summary

- Earth is made mostly of rock.

- All rocks come from **magma**.

- **Minerals** form when **magma** cools slowly, allowing the atoms to line up in an orderly fashion.

- **Dirt** is called **soil**. It is made mostly from rocks and contains materials that come from plants, **minerals**, and animals.

I love dirt!

15.5 Vocabulary

Basic Words

calcite   (KAAL-siyt)

magma   (MAAG-muh)

mica   (MIY-kuh)

mineral   (MIHN-ruhl)

quartz   (KWAWRTS)

Challenge Words

element   (EH-luh-muhnt)

igneous   (IHG-nee-uhs)

metamorphic   (meh-tuh-MAWR-fihk)

sedimentary   (seh-duh-MEHN-tuh-ree)

A rock with layers!

Our Earth

16.1 Shape of Earth

What is the shape of Earth? Is it round like a circle? Is it spherical like a ball? Is it flat like a pancake? Is it square like a block? Does it look like a pumpkin or an eggplant?

When you take a walk around your neighborhood, Earth seems flat. As you go to the park or walk to the store, you don't slide off the ground and you don't lean to the left or right. When you walk on a hill, you go up and then down again. You can walk for many miles and Earth will seem flat.

However, if you look at Earth from an airplane or from a boat at sea, you'll discover that Earth is not flat, but curved. If you looked at Earth from a spaceship, you would discover that Earth is shaped like a ball.

But Earth is not shaped like a perfectly round ball. The middle of Earth is pushed out a little, and the top and bottom are slightly flattened.

▶ Earth is shaped like a slightly smashed ball!

North Pole

The top part of Earth is called the **North Pole**, and the bottom part of Earth is called the **South Pole**. The middle of Earth is called the **equator**.

equator

South Pole

16.2 Size of Earth

It's hard to imagine that we walk, run, and live on a planet that's shaped like a ball. If you live at the middle of Earth, near the equator, do you feel like you are standing sideways? If you go to the South Pole do you feel upside down? If you go to the North Pole do you feel right-side up?

In fact, if you go anywhere on Earth, you don't feel upside down or sideways. You always feel right-side up.

Why does Earth seem flat if it is actually a curved ball? And why do we always feel upright even when we are at the South Pole or on the equator?

Many years ago some people were afraid to travel too far out to sea. They thought Earth was flat and that they would fall off the edge if they went too far! It made sense to them because Earth felt flat.

What they didn't know is that Earth is very large. Earth is so large that we can't feel Earth's curve when we walk or run. Earth is so large that we can't tell if we are on the top, on the bottom, or on the side.

Earth is about 40,200 kilometers (25,000 miles) around. It would take you more than a year to walk around Earth! That's how big Earth is.

▶ Because Earth is so large, it feels flat when we walk, run, work, or play. But Earth is really a huge ball.

16.3 Parts of Earth

Earth has different parts. You know that when you walk outside, you step on rocks and dirt. Rocks, minerals, and dirt make up the outer part of Earth. Rocks, minerals, and dirt also make up the ocean floor.

But what is beneath the dirt and the ocean? What is inside Earth? Is it rocks, minerals, and dirt like the outer part? Or is it chocolate syrup, liquid gold, or melted cheese?

Scientists have no way to actually see what's below the outer surface of Earth, but by studying volcanoes and earthquakes, scientists can come up with some ideas about what lies below the surface.

16.4 Earth's Layers

▶ Earth is a rock planet, which means that it is made mostly of rocks.

If scientists were able to cut Earth in half, they think they would see at least three different layers. This means that the outer surface of Earth, where you walk, is different from the part just below it.

The **crust** is the outermost layer of Earth. The crust is made up of rocks, minerals, and soil. The crust is very hard. This hard outer layer supports you when you walk and it supports buildings. The crust also holds the oceans. The crust is relatively thin compared to the layers below it, and it makes up only a small part of Earth.

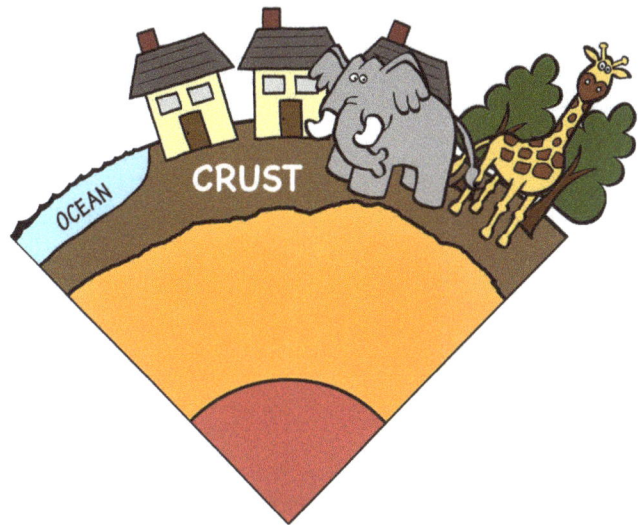

Below the crust is the **mantle**. Scientists think the mantle is much thicker than the crust and is made of layers. Scientists believe the outer part of the mantle is hard and rocky. The inner part of the mantle is thought to be softer and hotter. Scientists think this inner part of the mantle might be more like gooey peanut butter and made of melted rock called **magma**. Scientists think that the magma in the mantle does not always stay in the same place but moves around!

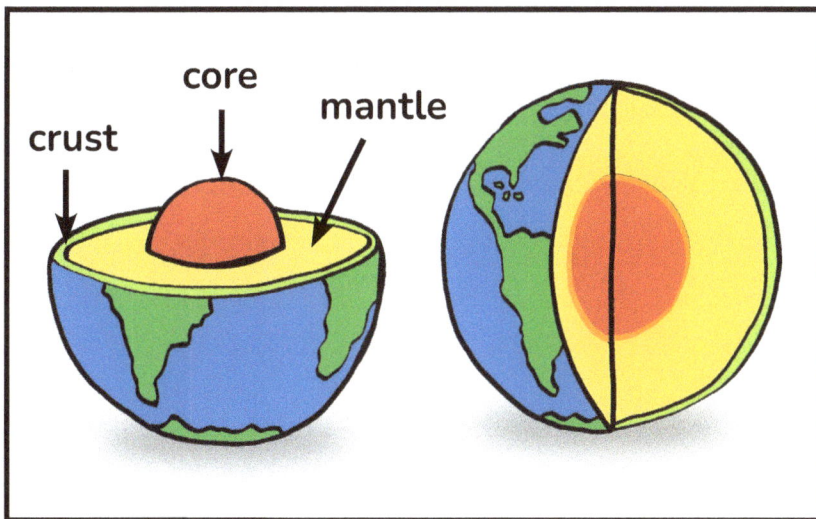

MANTLE

CRUST

CORE

Magma in the Mantle Moves!

Below the mantle and in the very center of Earth is the core. Scientists believe that the outer part of the core is made of molten iron and nickel and the inner part of the core is solid and made mostly of iron. The core is very heavy.

▶ Earth's crust, mantle, and core together are called the **geosphere**.

16.5 Summary

- Earth is made of layers.

- The main layers of Earth are the **crust**, the **mantle**, and the **core**.

- The **crust** is the hard, rocky, outermost layer of Earth.

- The **mantle** is below the **crust**. Most of the mantle is soft and the outer part closest to the **crust** is hard.

- The **core** is at the center of Earth. The inner part of the **core** is likely solid.

16.6 Vocabulary

Basic Words

mantle   (MAAN-tuhl)

Challenge Words

equator   (ih-KWAY-tuhr)

geosphere   (JEE-oh-sfihr)

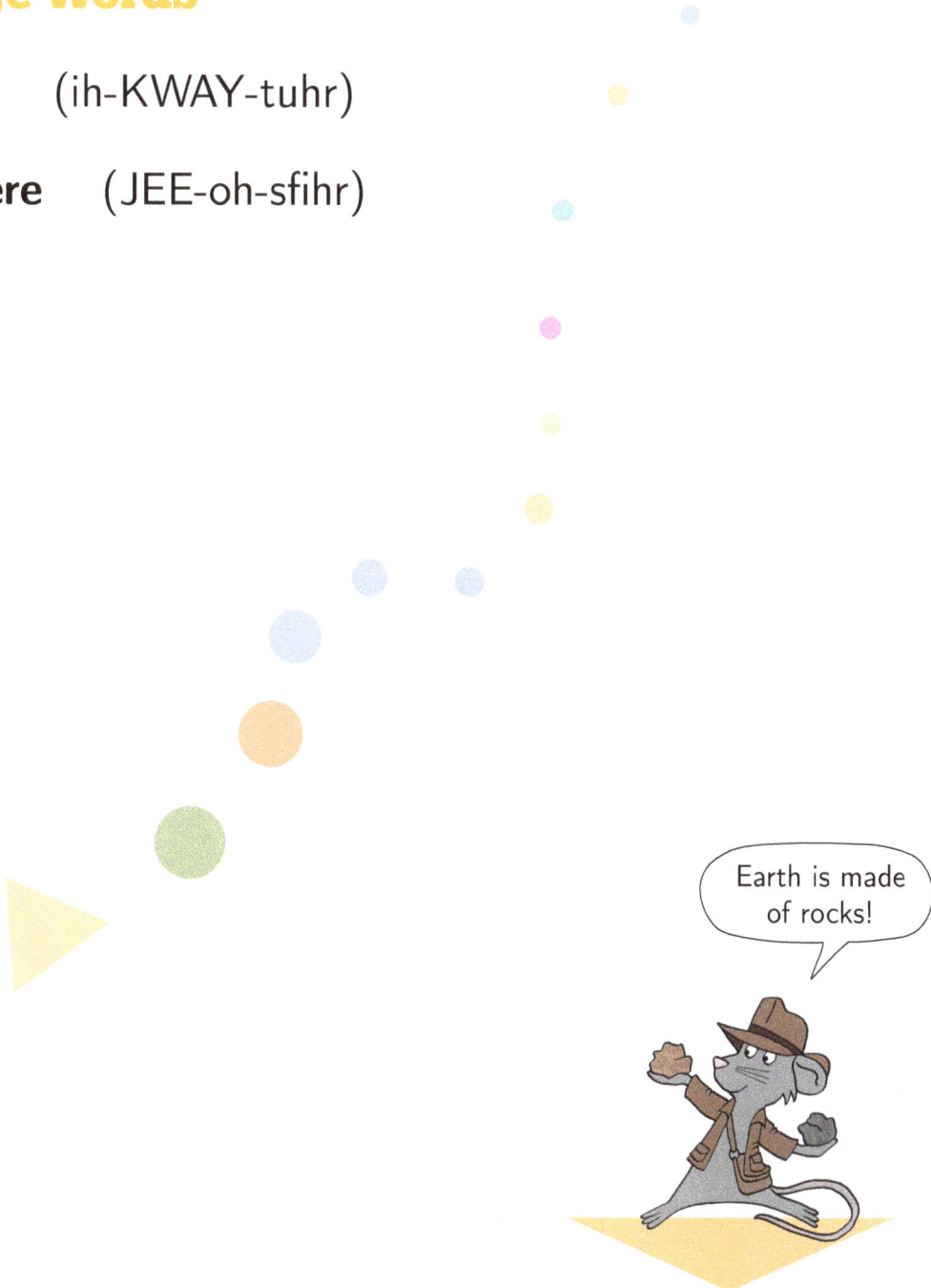

Earth is made of rocks!

Earth Is Active

17.1 Introduction

Earth is an active planet. It is always changing. For example, we can see the surface of Earth change with the seasons. In the spring and summer some parts of Earth can be covered with flowers and green grass. In the winter, these same parts might be covered with snow.

You might think that Earth's crust doesn't change because it is hard and rocky, but it changes every day. There are volcanoes, earthquakes, storms, and flowing rivers that continuously change Earth's crust.

Geologists study how Earth has changed by looking at rocks, dirt layers, mountains, and other features. Geologists also look at volcanoes, earthquakes, weather, and other things that cause changes on Earth.

17.2 Volcanoes Erupt

Volcanoes can be very exciting. Sometimes when a volcano erupts, a whole mountaintop will come off in a huge explosion.

▶ **Volcanoes** happen when magma in the mantle pushes up through a weak spot in the crust and comes to the surface of Earth.

▶ **Magma** is formed when the rocks and minerals in the mantle melt because the weight of the crust is pushing down on them, creating heat and pressure.

▶ **Pressure** happens when you squeeze something and it doesn't have anywhere to go.

For example, if you keep the lid on your toothpaste and squeeze the bottom of the tube, you will create pressure in the tube. If the top suddenly pops off, the toothpaste will explode! If there is a crack in the wall of the toothpaste tube, the toothpaste will squirt out the side.

This is what happens when a volcano erupts. The magma finds a crack or weak spot in the crust, and because of the pressure in the mantle, the molten rock will spew out.

When the magma comes out onto the surface of Earth, it is called **lava**. Lava that has flowed on Earth's surface is responsible for many interesting features that are created when the lava cools and forms hard rock. Sometimes **lava flows** form whole islands. The Hawaiian Islands, for example, are a group of islands made from lava flows.

Volcanoes can form mountains of different shapes and sizes. Some volcanoes form gently sloping mountains over long periods of time. These volcanoes are called **shield volcanoes** because their long, sloping sides make them resemble the shape of a shield. Shield volcanoes form when very thin layers of lava flow out in all directions. A mountain is formed as the layers build up on top of each other. Shield volcanoes can be several miles long with sides that slope very gradually. The Hawaiian Islands are a series of shield volcanoes.

Volcanoes can also form mountains with steeper sides. **Cone volcanoes** are cone-shaped because the magma spurts out more quickly and is thicker than the magma that forms shield volcanoes. Also, more rocks and dirt are scattered. The rocks and dirt pile up along the sides of the volcano, making the sides steeper and steeper. Interestingly, cone volcanoes are often found on the edges of shield volcanoes.

Volcanoes can also form dome-shaped mountains. Lava **dome mountains** are often round in shape and look like a cereal bowl turned upside down. The magma that comes from dome volcanoes is very thick and doesn't flow very far away from the center.

17.3 Earthquakes Shake

Earthquakes can also be very exciting and even scary if you happen to be near one when it happens. Earthquakes occur because sections of Earth's crust suddenly move. Recall that the hard, rocky crust lies on top of the hard outer part of the mantle, and the hard outer part of the mantle lies on top of the inner part of the mantle that is soft like peanut butter.

Scientists think that the outer part of the mantle is cracked into huge pieces called **plates** that fit together like a big puzzle. These pieces, or plates, float on the part of the mantle that is soft like peanut butter.

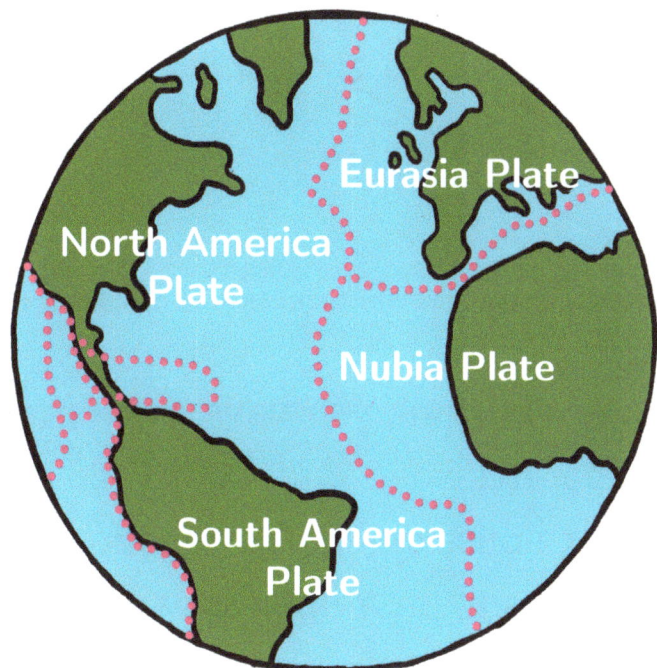

Earthquakes happen suddenly and are usually over within a few minutes. If the earthquake is small, you might feel the floor move, and you might hear a sound like a nearby train. If the earthquake is large, the whole ground moves and buildings and trees may fall. Sometimes after a big earthquake, several smaller earthquakes occur.

When plates move against each other, one section might move upward and the other might move downward. Or the sections might slide past one another. When these things happen, there is movement of the land on either side of the line that is between the two sections.

Let's imagine you have a street in front of your house that separates your yard from your neighbor's yard. Imagine that your house sits on one section of Earth and your neighbor's house sits on a different section of Earth. If these two sections move past one another, it's possible that your neighbor from across the street is now your next-door neighbor!

17.4 Summary

- Earth's surface is constantly changing.

- **Volcanoes** and **earthquakes** change Earth's crust.

- **Volcanoes** happen when magma in the mantle pushes up through a weak spot in Earth's surface.

- **Earthquakes** happen when sections of Earth's surface move.

17.5 Vocabulary

Basic Words

earthquake   (UHRTH-kwayk)

pressure   (PREH-shuhr)

lava flow   (LAH-vuh FLOH)

shield   (SHEELD)

Challenge Words

volcano   (vahl-KAY-noh)

Geology is fun to learn!

What Is ASTRONOMY?

18.1 Introduction

So far we have studied four important building blocks of science — chemistry, biology, physics, and geology. We'll now look at the fifth building block of science: **astronomy.**

Astronomy is the study of the cosmos. The term **cosmos** refers to Earth and everything that extends beyond Earth, including other planets, stars, nebulae, comets, asteroids, and even black holes.

18.2 Who Was the First Astronomer?

Because astronomers can't fly to faraway planets or ride asteroids, astronomers use various tools and techniques to find out more about the objects in the cosmos. However, before the use of modern tools, people could learn a great deal about the cosmos by studying the night sky.

It's hard to say who was the first astronomer. Many early people studied the planets and stars, and even without modern tools they discovered a great deal about the cosmos.

Early Egyptian, Babylonian, and Mayan people observed the sky in great detail. Noting when the Moon was full or when the Sun sank lower on the horizon, early observers were able to learn about how the planets and the Moon moved. From their observations they produced calendars and were able to predict eclipses.

▶ One of the questions early astronomers asked was, "Does Earth move around the Sun, or does the Sun move around Earth?"

In other words, do we live in a "Sun-centered" cosmos or an "Earth-centered" cosmos? To early astronomers it appeared from simple observation that we live in an Earth-centered cosmos. When the Sun rises and sets each day, it has the appearance of moving around Earth. However, as we will see, sometimes how things move isn't always easy to figure out.

▶ **Aristarchus of Samos**

One of the first astronomers to propose that Earth moves around the Sun was Aristarchus of Samos. Aristarchus was a Greek astronomer and mathematician who lived from 310-230 B.C.E. He studied the planets and said that Earth

Aristarchus 310-230 B.C.E.

has two different movements. One movement is that Earth travels around the Sun, and the other movement is that Earth revolves around its own axis. We now know that he was right! But during his time no one believed him. It would be almost 2000 years before astronomers would look closely at his ideas.

18.3 Famous Early Astronomers

▶ **Nicolaus Copernicus**

Nicolaus Copernicus was a famous astronomer who also thought that Earth moved around the Sun. Copernicus was born in 1473 C.E. in the ancient Polish city of Torun. During the time Copernicus lived, most scientists believed that the Sun revolved around Earth. They believed that Earth was the center of the universe and everything revolved around it.

Copernicus 1473-1543 C.E.

Copernicus did not agree with the scientists of his day. His ideas would eventually change the whole science of astronomy. Unlike Aristarchus, Copernicus was able to use mathematics to show that Earth moves around the Sun and that the Sun remains fixed in one location. However, Copernicus was not outspoken about his ideas. Because he knew his ideas might upset people, he didn't talk about them. When Copernicus did publish his work, a few people got upset, but most people just ignored his hard work. Another 100 years passed before people took his ideas seriously.

▶ Tycho Brahe

Another famous astronomer also changed the way we see the cosmos. His name was Tycho Brahe. Brahe was born in 1546 C.E. in the Danish town of Scania, and he was raised by his uncle. Like Copernicus, Brahe was curious about astronomy. His uncle wanted him to be a lawyer or a politician, but Brahe studied mathematics

Tycho Brahe 1546-1601 C.E.

and slipped away at night to look at the sky. When his uncle died, Brahe was free to pursue his interests in astronomy.

Telescopes that make faraway objects look closer were not yet invented, so Brahe used **sighting tubes**, which are just hollow tubes with no lenses. Sighting tubes can be used to look at one star at a time. In this way, Brahe discovered that stars do not always appear to be in the same position but are constantly changing. Based on his observations, Brahe decided to rewrite the map of the stars and spent his life working on his ideas.

▶ Galileo Galilei

Galileo Galilei was also a famous early astronomer. He was interested in trying to find out how the planets move. Galileo was born in 1564 C.E. in Pisa, Italy. He studied many different subjects, such as mathematics and physics, and he loved to look at the stars. Galileo used his knowledge of math and

Galileo Galilei 1564-1642 C.E.

physics to better understand how the planets and the Moon move.

Like Copernicus, Galileo was an independent thinker, and he didn't believe in an Earth-centered universe. Galileo did experiments because he wanted to show how things moved rather than just coming up with ideas about it. By doing

153

experiments and by using mathematics and physics, Galileo was able to **prove** that we live in a Sun-centered **solar system** that is made up of the Sun and the objects traveling around it. Being able to prove an idea by using experiments, math, and physics was the beginning of astronomy as a science.

18.4 Astronomers Today

Today, many scientists study the stars and planets. Astronomy is a science, and modern astronomers are scientists who use a variety of scientific tools and scientific techniques to learn about the universe.

However, even with new tools, modern astronomers must use the same basic skills that Copernicus, Brahe, and Galileo used.

Today's astronomers must make good observations and must train themselves to see the details, like Copernicus did. Astronomers must also study math and physics like Brahe and Galileo did.

Math and physics are essential for understanding how the stars and planets move in space. Most importantly, astronomers must always be curious and willing to argue to defend their ideas like Copernicus, Brahe, and Galileo did.

18.5 Summary

- **Astronomy** is the study of space and all the objects found in space.

- Early **astronomers** were able to discover a great deal about the stars and planets by using observation.

- **Nicolaus Copernicus**, **Tycho Brahe**, and **Galileo Galilei** were three early **astronomers** who changed the way we understand the universe.

- Modern **astronomers** still use observation, math, and physics to study space.

18.6 Vocabulary

Basic Words

cosmos   (KAHZ-muhs)

sighting tube   (SIY-tihng TOOB)

Challenge Words

Aristarchus of Samos   (aa-ruh-STAHR-kuhs
                             uhv SAY-mahs)

astronomy   (uh-STRAH-nuh-mee)

Nicolaus Copernicus   (NIH-koh-luhs koh-PUHR-nih-kuhs)

solar system   (SOH-luhr SIH-stuhm)

telescope   (TEH-luh-skohp)

Tycho Brahe   (TEE-koh BRAH-hee)

This telescope makes the stars look closer!

Earth's Home in SPACE

19.1 Introduction

Now that we know what astronomy is and how to study planets and stars, it's time to explore what Earth looks like from space.

Because Earth is so big compared to our human size, it's hard to imagine what Earth looks like from space. Is Earth the biggest object in space? Is Earth in the center of space? If we took a rocket into space, what would we find?

19.2 Earth Is a Planet

If you launched a rocket and traveled past the clouds into space, you would see Earth. Earth would look like a blue marble floating in the dark space around it.

▶ Earth is a **planet**.

A planet has special properties. A planet must move in space around a sun, must have enough **gravity** to have cleared other objects of a similar size out of its orbit, and must be **spherical** (shaped like a ball). **Gravity** is the force that holds everything to Earth's surface.

▶ Because Earth is very large, moves around the Sun in space, has gravity, and is spherical, Earth is called a planet.

Earth **rotates** around an **axis**, which is an imaginary line that goes through the center of an object. If you were to take a ball and spin it with your fingers, it would rotate around an axis.

▶ The Sun shines on different parts of Earth as Earth rotates. This is how we get our days and nights.

It takes 24 hours for Earth to rotate once around its axis. We don't feel the rotation of Earth because Earth's gravity holds down the air and everything else that's on Earth. Everything is moving at the same speed, so Earth's rotation does not cause wind.

▶ Earth is **tilted** on its axis. "Tilted" means that Earth is not just straight up and down but is slanted. The tilt of Earth's axis gives us **seasons**.

For part of the year, the northern part of Earth is tilted towards the Sun and the southern part of Earth is tilted away from the Sun. This gives the northern part of Earth summer and the southern part winter.

Then, during a different part of the year, the southern part of Earth is tilted towards the Sun and the northern part away from the Sun. When this happens, the southern part of Earth has summer, and the northern part has winter.

19.3 The Moon and Tides

If you take your rocket into space, you might run into the Moon. A **moon** is a natural object that travels around a planet. Our moon is smaller than Earth and travels around Earth.

Our moon does have some gravity and is spherical, but because it moves around Earth and not around the Sun as the Earth does, the Moon is not a planet. We will learn more about the Moon in the next chapter.

Did you know that the Moon helps create ocean tides on Earth? It's true! The Moon pulls on Earth with some gravity. This pulling on Earth causes the water in the oceans to be pulled too.

As the Moon moves around Earth, it pulls the ocean water with it. The pulling of ocean water by the Moon helps create **tides.**

19.4 The Sun and Weather

If you wear sandals during a summer day, you can feel the Sun warming your toes. The Sun is a big ball of fire that gives light and heat energy to Earth.

Earth orbits the Sun. The word **orbit** means to "revolve around." An orbit is the path one object makes as it travels around another object. If you stand with your hand on a pole and then start walking, you will make a path around the pole. You will orbit the pole. This is what Earth does as it moves around the Sun. We will learn more about the Sun in the next chapter.

Did you know there are storms on the Sun? Did you know that the storms on the Sun can cause storms on Earth? It's true! Sun storms can contribute to Earth storms. Scientists who research weather can study Sun storms to find out how they affect Earth's weather.

19.5 Eclipses

During the Moon's orbit around Earth, the Moon travels behind Earth. When Earth is in between the Moon and the Sun, the Earth can block the Sun's light from reaching the Moon. When the Earth's shadow is cast on the Moon, it is called a **lunar eclipse**.

At other times, the Moon will be in between the Sun and the Earth. With the Moon in this position, the Moon can block the Sun's light from reaching a portion of Earth. This is called a **solar eclipse**.

19.6 Summary

- Earth is a **planet**.

- One **rotation** of the Earth on its **axis** takes 24 hours (one day).

- Earth is **tilted** on its **axis**, giving us seasons.

- The Moon and Sun affect Earth's tides and weather.

19.7 Vocabulary

Basic Words

axis   (AAK-suhs)

lunar eclipse   (LOO-nuhr ih-KLIHPS)

orbit   (AWR-buht)

planet   (PLAA-nuht)

rotation   (roh-TAY-shuhn)

rotate   (ROH-tayt)

season   (SEE-zuhn)

solar eclipse   (SOH-luhr ih-KLIHPS)

Challenge Words

gravity   (GRAA-vuh-tee)

spherical   (SFIHR-ih-kuhl)

Wow! Earth turns to make day and night!

Earth's Neighbors
MOON & SUN

20.1 Introduction

In the last chapter you saw that Earth sits in space. You also saw that the Moon and Sun cause changes in our ocean tides and weather. But what is a moon and what is a sun?

20.2 The Moon

You can see the Moon from Earth. If you look outside your house, you might see a bright, round shape in the sky. This is the **Moon**.

You can sometimes see the Moon during the day, but most often you see the Moon as the brightest object in the sky at night.

▶ The Moon is spherical like Earth, but it is much smaller. The Moon has a very different surface from that of Earth. Although the Moon is made of rocks and minerals like Earth, the Moon cannot support life. It has very little oxygen and no liquid water.

The Moon looks bright in the sky, but the Moon does not generate its own light. Acting like a mirror, the Moon reflects the Sun's light to Earth.

The Moon orbits Earth once a month. If you look at the Moon often, you will see that it appears to change its shape during the month. The different shapes are caused by different views of the Moon as we see it from Earth. The Moon can be **round**, **half-round**, or **crescent** shaped. When the Moon looks round, we say it is a **full moon**.

If you look closely at the Moon, you can see light and dark patches. These light and dark patches sometimes look like faces. The "Man in the Moon" is a famous nursery rhyme from Mother Goose.

The light and dark patches on the Moon are actually **craters** and **lava flows**.

20.3 The Sun

The other large object you see in the sky is the **Sun**. You can't look directly at the Sun (that would damage your eyes), but you can see that the Sun is rising in the morning, moving across the sky during the day, and setting at night.

▶ The Sun is not a planet or a moon. The Sun is a **star**. A star is any object in space that generates its own light and heat energy (see Section 20.4).

The Sun is much larger than Earth. The Sun is so large that a million Earths would fit inside!

The Sun

The Sun is also very hot. The temperature on the Sun's surface is thousands of times hotter than temperatures on Earth. The Sun is so hot that it would melt everything on Earth if Earth were too close to it!

20.4 The Sun's Energy

▶ The Sun generates its own light and heat energy.

In Chapter 11 you learned that energy is something that gives something else the ability to do work. The energy made by the Sun comes out as light energy and heat energy. Plants use the Sun's light energy to make food. Making food is work, so the Sun gives plants the ability to do work. We can use the Sun's light energy to make electricity. Our bodies also use the Sun's heat energy to stay warm. Without the Sun there would be no life on Earth.

▶ The Sun is not made of rocks and is not like the Earth and the Moon. Instead, the Sun is made of helium and hydrogen gases. The extremely hot temperatures on the Sun make hydrogen atoms stick together, forming helium atoms.

In Chapter 4 you learned that when atoms stick together they can make molecules. You also saw in Chapter 5 how atoms and molecules rearrange to make new molecules in a chemical reaction. When atoms stick together to make molecules, they use their electrons to attach to each other.

▶ When hydrogen atoms stick together to make helium atoms, they rearrange their protons and neutrons. This is called a **nuclear reaction**.

A nuclear reaction is different from a **chemical reaction**. In a chemical reaction, the protons and neutrons stay the same and only the electrons stick together.

During a nuclear reaction lots of light and heat energy are released. Because the Sun can generate its own energy, it is called a star.

20.5 Summary

- The **Moon** orbits the Earth once a month.

- The **Moon** is made of rocks and minerals, like Earth.

- Our **Sun** is a **star** and makes its own energy.

20.6 Vocabulary

Basic Words

crater   (KRAY-tuhr)

crescent   (KREH-suhnt)

Challenge Words

nuclear reaction   (NOO-klee-uhr ree-AAK-shuhn)

I want to go to the Moon!

Earth's Neighbors
PLANETS

21.1 Introduction

In Chapter 20 we learned about two of Earth's neighbors — the Moon and the Sun. In this chapter we will learn about some of Earth's other neighbors — the **planets.**

21.2 Planets

If you look into the sky during the day, you can see the bright Sun and possibly a very faded Moon.

But at night, depending on where you live, the sky twinkles up with brilliant specks of light. Many of these specks are stars, like our Sun. But some of the bright lights that dot the night sky are planets.

In Chapter 19 we found out that Earth is a planet. Recall that Earth is a planet because it orbits the Sun, is spherical in shape, and is large enough to have cleared its orbit The Moon is not a planet because it orbits Earth rather than the Sun.

▶ Earth is one of eight planets that orbit the Sun. The names of these eight planets are: **Mercury**, **Venus**, **Earth**, **Mars**, **Jupiter**, **Saturn**, **Uranus**, and **Neptune**.

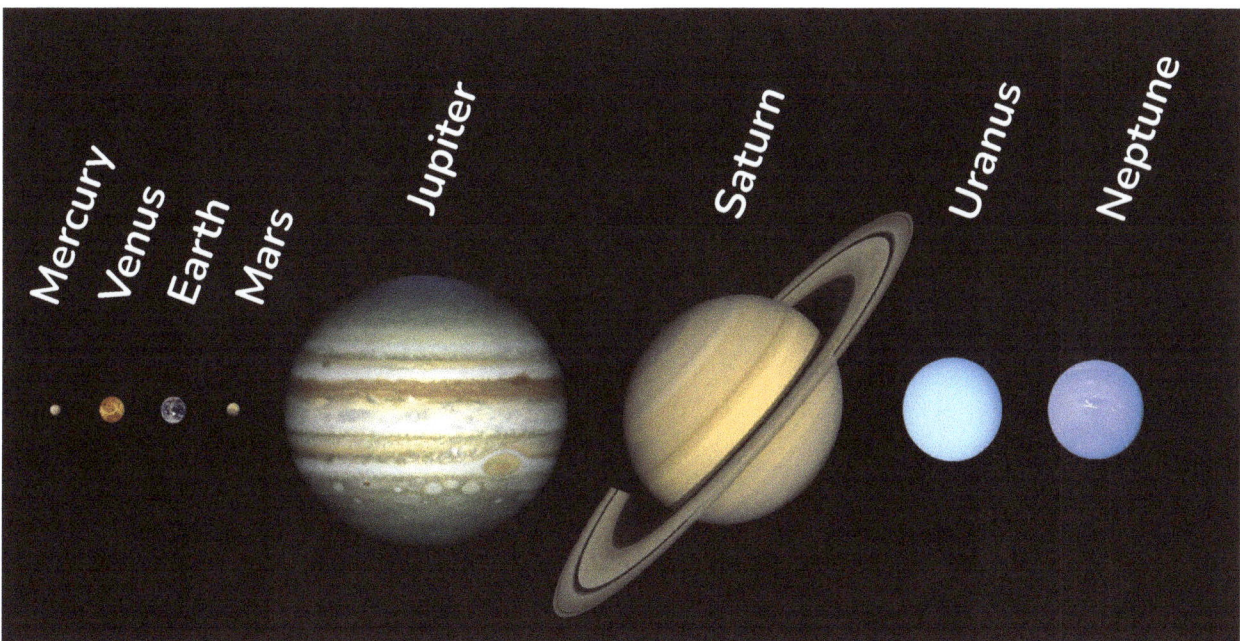

21.3 Two Types of Planets

All of the planets are different from each other. Mercury has a barren moon-like surface, Venus has toxic gas clouds, and Saturn has brilliant rings. Mars has a reddish color, and Uranus and Neptune look blue or blue-green.

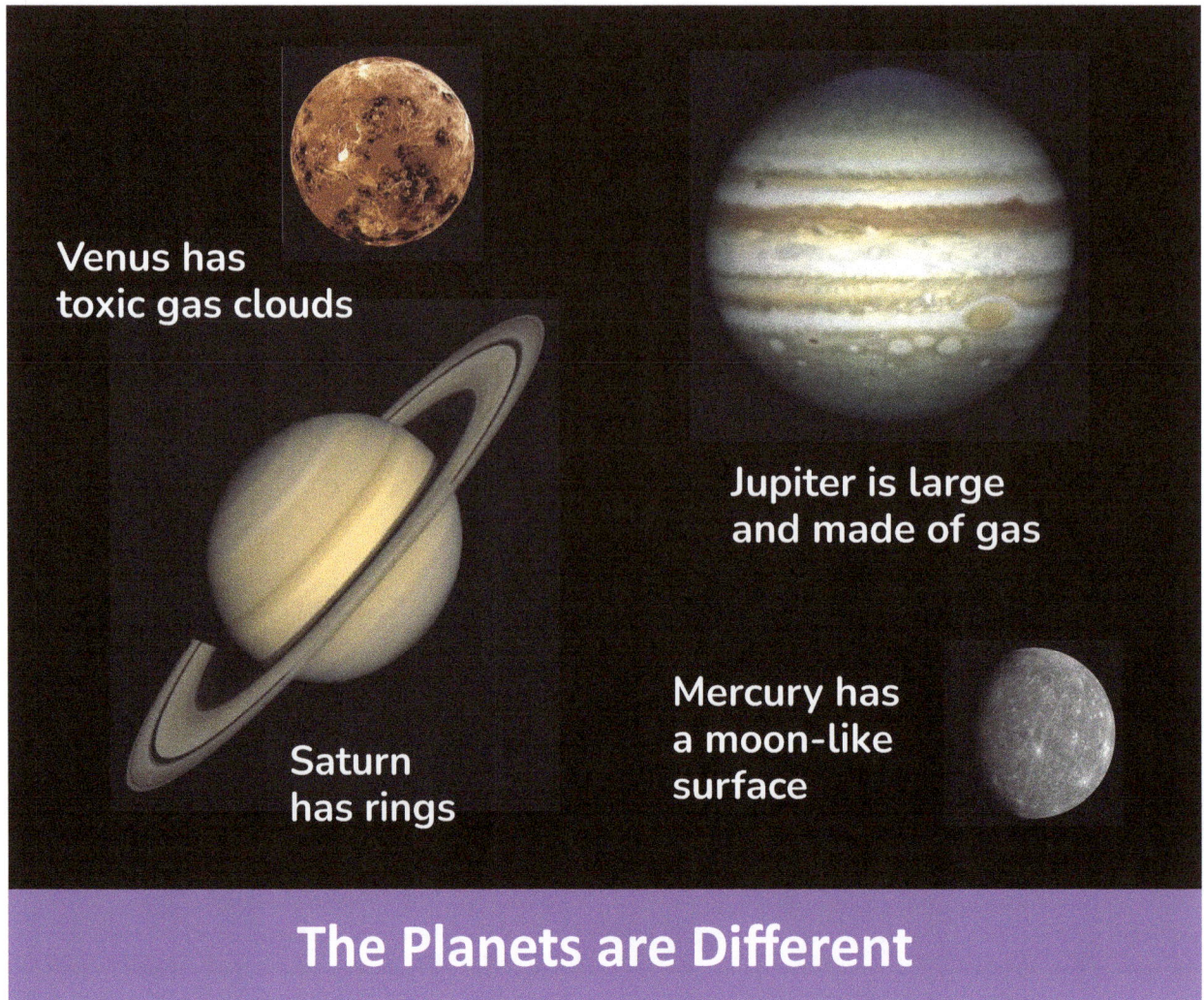

Venus has toxic gas clouds

Jupiter is large and made of gas

Saturn has rings

Mercury has a moon-like surface

The Planets are Different

Even though all the planets are different from each other, some of the planets have features that are similar. Because of these similarities, scientist are able to separate the planets into two groups. The names of these two groups are **terrestrial planets** and **Jovian planets**.

▶ The **terrestrial planets** are those planets that are most like Earth, and the word terrestrial means "Earth-like." There are four terrestrial planets: **Mercury, Venus, Earth,** and **Mars.**

All of the terrestrial planets are made of rock and minerals, like Earth. Also, all of the terrestrial planets have volcanoes, mountains, and craters on their surfaces.

Earth-like (terrestrial) Planets

However, of all the terrestrial planets, only Earth has life on it. Only Earth has the water, the oxygen, and the proper atmosphere needed to support life as we know it.

▶ The **Jovian planets** are those planets that are similar to Jupiter. Jupiter is a very large planet made mostly of hydrogen gas and helium gas. All of the Jovian planets are like Jupiter because they are all very large and made mostly of gas. The Jovian planets are **Jupiter, Saturn, Uranus,** and **Neptune.**

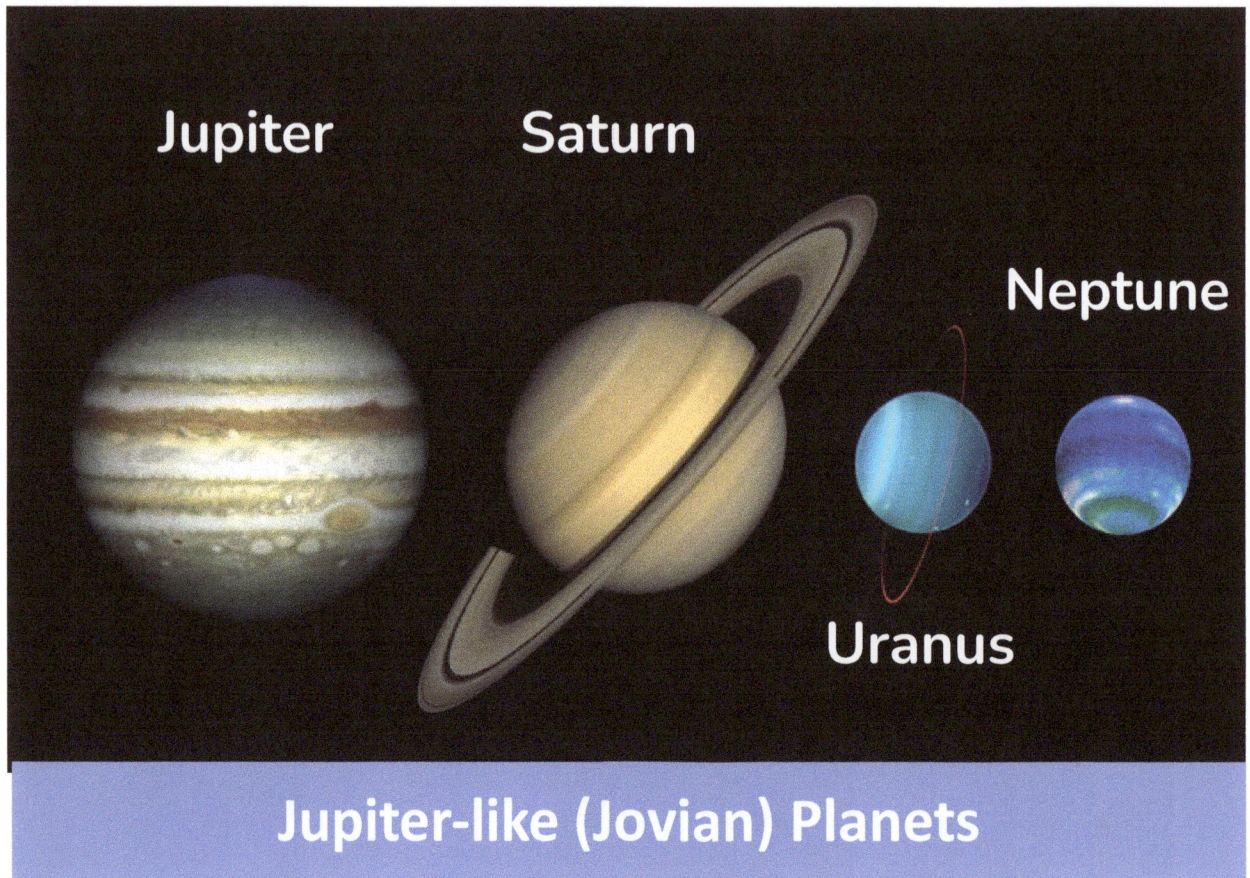

Jupiter

Saturn

Neptune

Uranus

Jupiter-like (Jovian) Planets

21.4 Where's Pluto?

Pluto was once called the 9th planet, but in August 2006 scientists at the **International Astronomical Union (IAU)** changed their minds. They decided Pluto does not have all the features needed to be classified as a planet. Pluto is now called a **dwarf planet** or a **plutoid**.

However, not all scientists agree with the IAU's decision. Scientists argue about conclusions all the time, and arguing is part of science. Someday Pluto may again be considered a planet.

21.5 Summary

- The eight planets are **Mercury**, **Venus**, **Earth**, **Mars**, **Jupiter**, **Saturn**, **Uranus**, and **Neptune**.

- The planets are separated into two groups: **terrestrial planets** and **Jovian planets**.

- **Terrestrial planets** are "like Earth." **Mercury**, **Venus**, **Earth**, and **Mars** are terrestrial planets.

- **Jovian planets** are "like Jupiter." **Jupiter**, **Saturn**, **Uranus**, and **Neptune** are Jovian planets.

I want to learn more about the planets!

21.6 Vocabulary

Basic Words

dwarf   (DWAWRF)

Earth   (UHRTH)

Mars   (MAHRZ)

Neptune   (NEHP-toon)

plutoid   (PLOO-toid)

Saturn   (SAA-tuhrn)

Venus   (VEE-nuhs)

Challenge Words

Jovian   (JOH-vee-uhn)

Jupiter   (JOO-puh-tuhr)

Mercury   (MEHR-kyuh-ree)

terrestrial   (tuh-REH-stree-uhl)

Uranus   (YOOR-uh-nuhs)

Putting It All
TOGETHER

22.1 Science

In this book you learned that **science** is a way to study and understand the world around us. You learned that **chemistry** is the study of atoms and molecules, and **physics** is the study of force, energy, and work. You learned that **biology** is the study of living things, **geology** is the study of Earth, and **astronomy** is the study of Earth in space, planets, and stars.

You learned that science gives us information about how things work, what they are made of, how they move, what they do, and whether or not they are alive.

22.2 All the Blocks Fit Together

Chemistry, biology, physics, geology, and astronomy are different parts of science. Like building blocks, they fit together to make science. Because they fit together, you have to learn a little about each building block to understand science as a whole.

▶ You will understand more chemistry when you learn about physics, biology, geology, and astronomy.

▶ You will understand more biology when you learn about chemistry, physics, astronomy, and geology.

▶ You will understand more physics when you learn about astronomy, geology, chemistry, and biology.

▶ You will understand more geology and astronomy when you learn about chemistry, physics, and biology.

The building blocks of science are connected to each other, and together they form the foundation for all science.

22.3 Learning Science

I WONDER IF THIS PLANET HAS THE RIGHT CONDITIONS FOR LIFE?

LET ME CHECK.

Chemistry

Learning about each of the different building blocks of science will help you understand science as a whole because all of the different science subjects fit together and overlap.

Chemistry explains not only how chemical reactions work in a test tube, but also how chemical reactions work in frogs, batteries, oceans, and the atmosphere. Chemistry also explains how nuclear reactions work in the Sun.

Physics explains not only how work, force, and energy take place in a marble, but also how work, force, and energy take place in butterflies and molecules and during earthquakes and solar storms.

Biology explains how living things live and breathe, and also helps geologists understand how living things change the Earth's surface. Biology helps astronomers guess whether or not life on other planets is possible.

LOOK WHAT I'VE DISCOVERED ABOUT YOU!

Geology explains what the Earth is made of and how changes to Earth affect living things. Geology uses physics to help us understand how volcanoes and earthquakes happen, and geologists also study other planets and moons to find out what they are made of and whether they might support life.

Astronomy helps us understand Earth's place in the cosmos. Astronomy uses chemistry and physics to discover what planets and stars are made of and how they move. Astronomers study how changes on Earth, the Moon, or the Sun affect the biology of life on Earth.

22.4 Discoveries in Science

If you decide to become a scientist, studying all the building blocks of science will be very important. You will have a much better chance of making new and exciting discoveries if you have knowledge of all the building blocks of science.

By studying each part of science you will be able to make connections you might otherwise miss. For example, by knowing chemistry, you might discover a new molecule important for butterflies. By knowing physics, you might find a new planet in a faraway galaxy. By knowing biology, you might discover a new chemical reaction important for frogs. By knowing geology, you might discover how life can adapt to the extreme environments on a glacier.

22.5 Summary

- **Science** is made of **five building blocks** — **chemistry**, **biology**, **physics**, **geology**, and **astronomy**.

- All of the **building blocks** of **science** fit together.

- Studying all the **building blocks** of **science** will help you learn more and will help you make new discoveries if you become a scientist.

22.6 Vocabulary

Review Words

astronomy   (uh-STRAH-nuh-mee)

biology   (biy-AH-luh-jee)

chemistry   (KEH-muh-stree)

geology   (jee-AH-luh-jee)

physics   (FIH-zihks)

science   (SIY-uhns)

I am going to make new discoveries!

Glossary / Index

(Pronunciation Key at end)

alcohol (AAL-kuh-hawl) • a type of molecule, 33

alive (uh-LIYV) • having life, 56

aluminum (uh-LOO-muh-nuhm) • an element found in some minerals, 125

animal (AA-nuh-muhl) • a living thing that has animal cells, 61

Animalia (aa-nuh-MAHL-yuh) • in taxonomy, the kingdom that contains all the animals, 61

Archaea (ahr-KEE-uh) • a domain that is one of the three major taxonomic divisions; also, the kingdom that contains the archaea organisms, 61

archaea (ahr-KEE-uh) [*singular, archaean* (ahr-KEE-uhn)] • a type of one-celled organism, 78

argon (AHR-gahn) • a type of atom; one of the noble gases, 46

Aristarchus of Samos (aa-ruh-STAHR-kuhs uhv SAY-mahs) • [310-230 B.C.E.] Greek astronomer and mathematician; studied the planets and said that the Earth revolves on an axis and travels around the Sun, 150

Aristotle (EHR-uh-stah-tuhl) • [384-322 B.C.E.] Greek philosopher; said to be the first biologist; made many observations of plants and animals; also studied motion, 50, 87

astronomy (uh-STRAH-nuh-mee) • the study of the cosmos, 6, 148, 179

atom (AA-tuhm) • the smallest unit of matter, 17, 21, 68, 123, 169

axis (AAK-suhs) [*plural,* **axes** (AAK-seez)] • an imaginary line that goes through the center of an object, 158

Bacteria (baak-TIHR-ee-uh) • a domain that is one of the three major taxonomic divisions; also, the kingdom that contains the bacteria, 61

bacteria (baak-TIHR-ee-uh) [*singular,* **bacterium** (baak-TIHR-ee-uhm)] • a type of one-celled organism, 77

beryllium (buh-RIH-lee-uhm) • an element found in some minerals, 125

biologist (biy-AH-luh-jihst) • a scientist who studies life, 49

biology (biy-AH-luh-jee) • the study of life, 6, 49, 179

Brahe, Tycho (BRAH-hee, TEE-koh) • [1546-1601 C.E.] Danish astronomer; made accurate observations of stars and planets; rewrote a star map, 152

calcite (KAAL-siyt) • a mineral found in rocks, 125

carbohydrate (kahr-boh-HIY-drayt) • a molecule made of carbon, hydrogen, and oxygen that gives energy to the body, 36, 100

carbon (KAHR-buhn) • a type of atom, 25, 38, 123

cell (SEHL) • the basic building block of living things, 67

chemical energy • see energy, chemical

chemical process (KEH-muh-kuhl PRAH-suhs) • any activity that involves chemistry, 14

chemical reaction (KEH-muh-kuhl ree-AAK-shuhn) • occurs when atoms join together, leave a molecule, or switch places, 44, 170

chemical stored energy • see energy, chemical stored

chemistry (KEH-muh-stree) • the investigation of what physical things are made of and the ways in which they change; the study of matter, 6, 13, 179

chlorine (KLAW-reen) • a type of atom, 25, 34

coal (KOHL) • a source of energy that comes from the Earth, 110

colony (KAH-luh-nee) • a large group of individuals living very close together, 80

conclusion (kuhn-KLOO-zhuhn) • a judgment that follows from information collected, 10

cone volcano • see volcano, cone

convert (kuhn-VUHRT) • to change from one form to another, 106

Copernicus, Nicolaus (koh-PUHR-nih-kuhs, NIH-koh-luhs) • [1473-1543 C.E.] Polish astronomer; thought the Earth moved around the Sun, 151

core (KAWR) • the very center part of Earth, 137

cosmos (KAHZ-muhs) • the Earth and everything that extends beyond the Earth, including other planets, stars, nebulae, comets, asteroids, and black holes, 148

crater (KRAY-tuhr) • a bowl-shaped depression around the opening of a volcano; a hole in the ground from a meteorite impact, 167

crescent (KREH-suhnt) • the shape of the Moon when less than half of it is visible, 167

crust (KRUHST) • in geology, the outer layer of Earth, 122, 137

crystal (KRIH-stuhl) • a solid that has a regular internal arrangement of atoms, 34, 125

dirt • soil; made from rocks, plants, minerals, and dead animals, 128

DNA (DEE-EHN-AY) • a molecule that holds information for the cell, 73

domain (doh-MAYN) • in taxonomy, one of the three largest groups of living things, 61

dome mountain (DOHM MOWN-tuhn) • a rounded volcano formed from thick lava, 144

dwarf planet • see planet, dwarf

Earth (UHRTH) • the planet we live on, 120, 157

earthquake (UHRTH-kwayk) • a shaking of Earth that happens when sections of the crust move suddenly, 145

eclipse, lunar (ih-KLIHPS, LOO-nuhr) • a darkening of the Moon as the Moon passes behind the Earth and the Earth's shadow falls on the Moon, 163

eclipse, solar (ih-KLIHPS, SOH-luhr) • a darkening of the Sun when the Moon passes between the Earth and the Sun and the Sun's rays are blocked from reaching Earth, 163

elastic stored energy • see energy, elastic stored

electron (ih-LEHK-trahn) • one of the three basic parts of an atom; found on the outside of the atom, 27

element (EH-luh-muhnt) • an atom, 123

emerald (EHM-ruhld) • a green mineral used in jewelry, 125

energy (EHN-uhr-jee) • something that gives something else the ability to do work, 93, 105

energy, chemical (EHN-uhr-jee, KEH-mih-kuhl) • energy produced by chemical reactions, 101, 106

energy, chemical stored (EHN-uhr-jee KEH-mih-kuhl STAWRD) • stored energy released by chemical reactions, 101

energy, elastic stored (EHN-uhr-jee, ih-LAAS-tihk STAWRD) • energy that is found in materials that can stretch, 101

energy, gravitational stored (EHN-uhr-jee, graa-vuh-TAY-shuh-nuhl STAWRD) • energy stored in objects that are elevated above the ground and can be pulled down by gravity, 101

energy, kinetic (EHN-uhr-jee, kih-NEH-tihk) • energy found in moving objects, 103

energy, mechanical (EHN-uhr-jee, mih-KAA-nih-kuhl) • energy having to do with the motion or position of an object, 106

energy, nuclear (EHN-uhr-jee, NOO-klee-uhr) • energy produced by nuclear reactions, 110

energy, stored (EHN-uhr-jee, STAWRD) • unused energy, 99

environmental geologist • see geologist, environmental

equator (ih-KWAY-tuhr) • the part of Earth halfway between the North Pole and the South Pole, 133

Eukarya (yoo-KAA-ree-uh) • one of the three major taxonomic domains; contains plants and animals, 61

experiment (ihk-SPEHR-uh-muhnt) • a test of ideas to observe what happens, 2, 9, 15, 87, 153

explosion (ihk-SPLOH-zuhn) • a big, fast, violent reaction, 47

feature (FEE-chuhr) • in biology, something that can be noticed and that helps describe a living thing, such as hair, hooves, feathers, or leaves, 60

Felis catus (FEE-luhs KAY-tuhs) • the scientific name for the house cat, 65

force (FAWRSS) • any action that changes the location of an object, the shape of an object, or how fast or slowly an object is moving, 93

fossil (FAH-suhl) • extremely old remains of a plant or animal that have been left in Earth's crust, 111

full moon • the phase of the Moon during which the Moon looks like a full, round circle, 167

Fungi (FUHN-jiy) • in taxonomy, the kingdom that includes mushrooms, molds, and yeasts, 61

fungus (FUHN-guhs) [plural, **fungi** (FUHN-jiy)] • a member of the kingdom Fungi, 64

Galileo Galilei (gaa-luh-LAY-oh gaa-luh-LAY) • [1564-1642 C.E.] Italian astronomer; used physics to study how planets, the Moon, and other objects move; proved the solar system is Sun-centered, 87, 153

gasoline (GAAS-uh-leen) • a source of energy that comes from the Earth; a fossil fuel, 44

genus (JEE-nuhs) [*plural,* **genera** (JEH-nuh-ruh)] • in taxonomy, the first part of the scientific name of a living thing, 64

geochemist (jee-oh-KEH-mihst) • a geologist who studies the chemistry of Earth, 116

geologist (jee-AH-luh-jihst) • a scientist who studies the Earth, 116

geologist, environmental (jee-AH-luh-jihst, ihn-viy-ruhn-MEHN-tuhl) • a geologist who studies Earth's environment, 117

geologist, structural (jee-AH-luh-jihst, STRUHK-chuh-ruhl) • a geologist who studies how Earth is put together and how it changes, 117

geology (jee-AH-luh-jee) • the study of Earth, 6, 113, 179

geosphere (JEE-oh-sfir) • Earth's crust, mantle, and core together; made of rock, minerals, and soils, 138

glucose (GLOO-kohs) • a type of sugar, 36

Golgi apparatus (GOHL-jee aa-puh-RAA-tuhs) • a part of a cell where proteins are made, 74

gravitational stored energy • see energy, gravitational stored

gravity (GRAA-vuh-tee) • the force that holds everything to Earth's surface, 89, 158

helium (HEE-lee-um) • a type of atom; one of the noble gases, 46

Homo sapiens (hoh-moh SAY-pee-ehnz) • the scientific name for humans, 65

hydrogen (HIY-druh-juhn) • a type of atom, 25, 35, 37

hydroxide (hiy-DRAHK-siyd) • a type of molecule, 41

hypothesis (hiy-PAH-thuh-sihs) • a statement that sets forth an assumption or idea about something, 9

igneous (IHG-nee-uhs) • a type of rock formed when magma inside the Earth cools and hardens, 126

International Astronomical Union [IAU] (ihn-tuhr-NAA-shuh-nuhl aa-struh-NAH-mih-kuhl YOON-yuhn) • an organization dedicated to advancing astronomy, 176

Jovian (JOH-vee-uhn) • Jupiter-like, 174

Jovian planet • see planet, Jovian

Jupiter (JOO-puh-tuhr) • a large, gaseous planet in Earth's solar system; fifth from the Sun, 173

kinesin (kih-NEE-suhn) • a "truck" that moves proteins around within a cell, 74

kinetic energy • see energy, kinetic

kingdom (KIHNG-duhm) • a division of a domain, 61

lava (LAH-vuh) • magma that has come out to the surface of the Earth, 143

lava flow (LAH-vuh FLOH) • molten or hardened lava around the mouth of a volcano, 143, 167

Linneaus, Carolus (lih-NEE-uhs, KAA-ruh-luhs) • [1707-1778 C.E.] Swedish physician considered to be the founder of taxonomy; came up with a system for sorting living things, 60

lunar eclipse • see eclipse, lunar

magma (MAAG-muh) • molten rock found deep inside Earth, 124, 137, 142

mantle (MAAN-tuhl) • the layer of Earth below the crust, 137

Mars (MAHRZ) • a terrestrial planet in our solar system, the fourth from the Sun, 173

matter (MAA-tuhr) • everything that can be seen, smelled, or touched, 13

mechanical energy • see energy, mechanical

Mercury (MEHR-kyuh-ree) • a terrestrial planet in our solar system; the closest planet to the Sun, 173

metamorphic (meh-tuh-MAWR-fihk) • a rock that has been changed from one type to another by means of very high heat and pressure, 126

method (MEH-thuhd) • a plan, procedure, or process that is followed, 8

mica (MIY-kuh) • a type of mineral that is soft and made of thin layers, 124

microscope (MIY-kruh-skohp) • an instrument that makes tiny things look bigger, 78

microtubule (miy-kroh-TOO-byool) • a long protein tube used as a "road" to move proteins around in a cell, 74

mineral (MIHN-ruhl) • a material formed by the slow cooling of magma; has atoms that are lined up in an orderly fashion, 122

model (MAH-duhl) • an object made from different materials, writing, drawings, or a computer program and used to explore an idea about how something works, 17, 26

molecule (MAH-lih-kyool) • two or more atoms hooked together, 33, 40

moon • a natural object that travels around a planet, 161, 165

natural gas (NAA-chuh-ruhl GAAS) • a source of energy that comes from the Earth; a fossil fuel, 110

neon (NEE-ahn) • a type of atom; one of the noble gases, 46

Neptune (NEHP-toon) • a Jovian planet in the Earth's solar system; eighth from the Sun, 173

neutron (NOO-trahn) • one of the 3 basic parts of an atom; found in the center of the atom, 27

Newton, Isaac (NOO-tuhn, IY-zihk) • [1642-1726 C.E.] English scientist and mathematician who used math to figure out many rules of physics, 88

nitrogen (NIY-truh-juhn) • a type of atom, 25, 38

noble gas (NOH-buhl GAAS) • a type of atom that doesn't usually react with other atoms or molecules, 46

nonliving thing (nahn-LIHV-ihng THIHNG) • an object that is not alive, 57

North Pole (NAWRTH POHL) • the very top point of Earth, 133

nuclear energy • see energy, nuclear

nuclear reaction (NOO-klee-uhr, ree-AAK-shuhn) • occurs when the protons and neutrons of an atom are rearranged, creating energy and changing the structure of the atom, 170

nucleus (NOO-klee-uhs) • the central part of a plant or animal cell; holds the DNA, 73

observation (ahb-suhr-VAY-shuhn) • the result of noticing a fact or facts about something, 1, 8, 50, 89

orbit (AWR-buht) • the curved path of a celestial body as it travels around another celestial body, 162

organism (AWR-guh-nih-zuhm) • a living thing, 78

oxygen (AHK-sih-juhn) • a type of atom, 25, 35, 38, 123

particle (PAHR-tih-kuhl) • in a chemical reaction, a small, solid piece of matter that results from the reaction, 47

phosphorus (FAHS-fuh-ruhs) • a type of atom, 25

physicist (FIH-zuh-sihst) • a scientist who studies physics, 86 physics (FIH-zihks) • the branch of science that explores how the natural world works, 6, 85, 179

planet (PLAA-nuht) • a large spherical celestial body that orbits a sun and clears its orbit, 158, 172

planet, dwarf (PLAA-nuht, DWAWRF) • a celestial body that does not have all the features needed to be classified as a planet; also called a plutoid, 177

planet, Jovian (PLAA-nuht, JOH-vee-uhn) • a large planet that is similar to Jupiter; also called a gas giant; Jupiter, Saturn, Uranus, and Neptune, 174

planet, terrestrial (PLAA-nuht, tuh-REHS-tree-uhl) • a planet that is Earth-like; made of rocky materials and closer to our Sun; Mercury, Venus, Earth and Mars, 174

plant (PLAANT) • a living thing in the kingdom Plantae, 61

Plantae (PLAAN-tiy) • in taxonomy, the kingdom that contains all organisms that have plant cells, 61

plate (PLAYT) • in geology, a huge piece of the outer part of the mantle, 145

Pluto (PLOO-toh) • once considered the 9th planet, is now classified as a dwarf planet or plutoid, 176

plutoid (PLOO-toid) • a celestial body that does not have all the features needed to be classified as a planet; also called a dwarf planet, 177

plutonium (ploo-TOH-nee-uhm) • an element used to create nuclear energy, 110

pressure (PREH-shuhr) • a force that moves things, 142

protein (PROH-teen) • a complicated molecule that does a job inside a cell in a living thing, 37

protist (PROH-tihst) • a tiny living thing in the kingdom Protista; also called a protozoan, 62

Protista (proh-TEE-stuh) • in taxonomy, the kingdom that contains protists, 61

proton (PROH-tahn) • one of the 3 basic parts of an atom; found in the center of the atom, 27

protozoa (proh-tuh-AOH-uh) • a protist, 63

prove (PROOV) • to show the truth of something, 154

quartz (KWAWRTS) • a common mineral found in rocks, 124

react (ree-AAKT) • in chemistry, when molecules meet and join together or break apart, 41

reaction (ree-AAK-shuhn) • a chemical reaction, 44

reproduce (ree-pruh-DOOSS) • in biology, for a living thing to make another living thing that is like itself, 57

ribosome (RIY-buh-sohm) • a protein used by a cell to make other proteins, 74

telescope (TEH-luh-skohp) • an instrument that makes far away objects appear to be closer, 152

temperature (TEHM-puhr-chuhr) • a measure of heat, 47

terrestrial (tuh-REH-stree-uhl) • Earth-like, 174

terrestrial planet • see planet, terrestrial

Theophrastus (thee-uh-FRAA-stuhs) • [371-287 B.C.E.] Greek philosopher; studied rocks, 115

tide (tiyd) • the rise and fall of ocean levels caused by the Moon and Sun, 161

tilt (TIHLT) • slant, 159

uranium (yoo-RAY-nee-uhm) • a type of atom, 29

Uranus (YOOR-uh-nuhs) • a Jovian planet, seventh from the Sun in Earth's solar system, 173

Venus (VEE-nuhs) • a terrestrial planet, the second from the Sun in Earth's solar system, 173

virus (VIY-ruhs) • a tiny bag of molecules that does not have all the parts of a complete cell; can cause diseases, 78

volcano (vahl-KAY-noh) • an eruption of molten rock caused by pressure forcing magma through a weak spot in the crust, 141

volcano, cone (vahl-KAY-noh, KOHN) • a steep-sided, cone-shaped volcano that forms quickly, 144

volcano, shield (vahl-KAY-noh, SHEELD) • a volcano with long, sloping sides formed over a long period of time, 143

weather (WEH-thuhr) • how hot or cold, windy or calm, wet or dry, clear or cloudy it is outside, 162

work (WEHRK) • in physics, an event that occurs when a force moves an object, 93

Pronunciation Key

aa	past	(PAAST)
ay	date	(DAYT)
ah	far	(FAHR)
aw	saw	(SAW)
eh	pet	(PEHT)
ee	feed	(FEED)
ih	trip	(TRIHP)
iy	side	(SIYD)
oh	bone	(BOHN)
oi	boy	(BOI)
oo	rule	(ROOL)
ow	noun	(NOWN)
uh	fun	(FUHN)
uu	wood	(WUUD)
y	yet	(YEHT)

Image Credits

Cover and Title Page: By yanosh nemesh, AdobeStock

Chapter 1: Adobe Stock: Page 1. By Liudmila; Page 4. By leelook

Chapter 2: Adobe Stock: Page 15. By Ivelin Radkov; Page 18. Top–By didesign; Bottom–By SeventyFour; Page 19. Top–By manusapon; Bottom–By SeventyFour

Chapter 3: Adobe Stock: Page 21. Top–By Roman Ivaschenko; Page 22. Top–By Hyrma; Middle–By oleh11; Bottom–By Contributor Artist; Page 23. Rock, By Nik Merkulov; Page 24. Clouds, By Krit; Carrots, By Hyrma; Peas, By oleh11; Brussels sprouts, By Nik Merkulov

Chapter 4: Adobe Stock: Page 32. By Anterovium; Page 33. Glass, By Anterovium; Page 34. By Pawel Burgiel, Page 36. By Dr Microbe

Chapter 5: Adobe Stock: Page 44. By Pathompong; Page 45. Top–By Shelby; Bottom–By MrP; Page 46. Illustration By TarikVision; Page 47. By Aleksey

Chapter 6: Page 53. Top–By Prrrettty, Adobe Stock; Bottom–By Sven Backstroem, Adobe Stock; Page 54. Top–By Heru Agung, iStockx; Bottom–By simonidadj, Adobe Stock

Chapter 7: Adobe Stock: Pages 58 & 59 By Mohd Khairil; Page 60. Photo By Matthias; Page 62. Plants By myboys.me; Snake By kuritafsheen; Bacteria By Steve Young; Page 63. Ameoba By micro photo; Mushrooms By avdwolde; Page 64. Photo By cynoclub

Chapter 8: Page 69. Bone Cell, Public Domain; Muscle Cells, Public Domain; Skin Cells By Dr_Microbe, Adobe Stock

Chapter 9: Adobe Stock: Page 80. By Artur; Page 81. Top–By kelly marken; Bottom–By Axel Kock; Page 82. Top–By Dr_Microbe; Bottom–By klyaksun; Page 83. By hpbfotos

Chapter 10: Page 86. By Sergey Novikov, Adobe Stock; Page 89. Photo by Meg Boulden on Unsplash

Chapter 11: Adobe Stock: Page 95. By emanelda; Page 97. By mamahoohooba

Chapter 12: Page 99. By triocean, AdobeStock

Chapter 13: Adobe Stock: Page 105. By teen00000; Page 106. By Mat Hayward

Chapter 14: Adobe Stock: Page 115. By olenap; Page 118. By Carlos

Chapter 15: Page 120. By alicja neumiler, AdobeStock; Page 125. Mica, By Pascal Terjan from London, United Kingdom, CC BY SA 2.0; Calcite, By Rob Lavinsky, iRocks.com, CC BY SA 3.0; Ruby, Public Domain; Page 127. Top–By vvoe, AdobeStock; Bottom–By Wilson, Public Domain; Page 128. Photo by Andres Siimon on Unsplash

Chapter 16: Page 132. NASA, Public Domain

Chapter 17: Adobe stock: Page 141. Top–By puckillustrations; Bottom–By bierchen; Page 143. By PhotoImage; Page 144. Top–By Jože Potrebuješ; Bottom–By K I Photography

Chapter 18: Page 154. By astrosystem, AdobeStock

Chapter 19: Page 158. NASA, Public Domain

Chapter 20: Page 166. By Delphotostock, NASA Elements, AdobeStock; Page 168. NASA

Chapter 21: Page 173-176. NASA

www.ingramcontent.com/pod-product-compliance
Lightning Source LLC
Chambersburg PA
CBHW040141200326
41458CB00025B/6340